SAFETY AND RISK IN SOCIETY

AMMONIUM NITRATE FACILITIES

SAFETY AND OVERSIGHT

SAFETY AND RISK IN SOCIETY

Additional books in this series can be found on Nova's website under the Series tab.

Additional e-books in this series can be found on Nova's website under the e-book tab.

SAFETY AND RISK IN SOCIETY

AMMONIUM NITRATE FACILITIES

SAFETY AND OVERSIGHT

ALLEGRA WARD
EDITOR

nova publishers
New York

For permission to use material from this book please contact us:
Telephone 631-231-7269; Fax 631-231-8175
Web Site: http://www.novapublishers.com

Library of Congress Cataloging-in-Publication Data

ISBN: 978-1-63321-716-4

Published by Nova Science Publishers, Inc. † New York

Contents

PREFACE

Chapter 1 – In April 2013, about 30 tons of ammonium nitrate fertilizer detonated during a fire at a facility in West, Texas, killing at least 14 people and damaging nearby schools, homes, and a nursing home. This incident raised concerns about the risks posed by similar facilities across the country. OSHA and EPA play a central role in protecting workers and communities from chemical accidents, and DHS administers a chemical facility security program. GAO was asked to examine oversight of ammonium nitrate facilities in the United States and other countries. This report addresses (1) how many facilities have ammonium nitrate in the United States, (2) how OSHA and EPA regulate and oversee facilities that have ammonium nitrate, and (3) what approaches selected other countries have adopted for regulating and overseeing facilities with ammonium nitrate. GAO analyzed available federal data and data from selected states with high use of ammonium nitrate; reviewed federal laws and regulations; and interviewed government officials, chemical safety experts, and industry representatives in the United States and selected countries.

Chapter 2 – The explosion on April 17, 2013, at the West Fertilizer Company fertilizer distribution facility in West, TX, has led to questions about the oversight and regulation of agricultural fertilizer. Facilities holding chemicals must comply with regulations attempting to ensure occupational safety, environmental protection, and homeland security. In addition to federal regulation requiring reporting and planning for ammonium nitrate and anhydrous ammonia, most state and some local governments have laws and regulations regarding the handling of either or both of these chemicals.

The West Fertilizer Company possessed a variety of agricultural chemicals at its retail facility, but policy interest has focused on two chemicals: ammonium nitrate and anhydrous ammonia. Ammonium nitrate is a solid that can be used as a fertilizer, a use that generally occurs without incident. In combination with a fuel source and certain conditions, such as added heat or shock, confinement, or contamination, ammonium nitrate can pose an explosion hazard. Such accidents have rarely occurred, but have historically had high impacts. For example, the ammonium nitrate explosion in 1947 in Texas City, TX, where two ships carrying ammonium nitrate coated in wax and stored in paper bags caught fire and exploded, destroyed the entire dock area, including numerous oil tanks, dwellings, and business buildings. The bomb used in 1995 to attack the Murrah Federal Building in Oklahoma City, OK, contained ammonium nitrate as a component of its explosives.

Chapter 3 – The Environmental Protection Agency (EPA), the Occupational Safety and Health Administration (OSHA), and the Bureau of Alcohol, Tobacco, Firearms, and Explosives (ATF) ("we") are issuing this advisory[1] as part of an ongoing federal effort to improve chemical risk management, and to advance safety and protect human health and the environment. This advisory contains information on recent and past accidents involving AMMONIUM NITRATE (commonly referred to as AN), on the hazards of AN, how to manage these hazards, and appropriate steps for community emergency planning and proper emergency response. It is focused primarily on safe handling and storage of higher density, solid AN pellets and prills (a prill is a small bead) used in fertilizers. This advisory is intended to broadly disseminate lessons learned from recent incidents involving AN so that such incidents can be prevented in the future. Also provided is a list of information resources, including relevant codes and standards, industry publications, and applicable statutes and regulations that will help facilities handling AN and first responders better understand the hazards so they can effectively manage the risks. The information provided is not intended to cover all the hazards, safe practices or technical challenges associated with the manufacturing of AN; liquid fertilizers containing AN; manufacturing, storage or use of explosives or blasting agents containing AN; or the transportation of AN. For these particular situations, please consult other sources including the appropriate references, standards and regulations, cited at the end of this document.

Chapter 4 – Testimony of Rafael Moure-Eraso, Chairman, U.S. Chemical Safety Board.

Chapter 5 – Testimony of Barry N. Breen, Principal Deputy Assistant Adminstrator, Office of Solid Waste and Emergency Response, U.S. Environmental Protection Agency.

Chapter 6 – Testimony of Paul Orum, Consultant, Coalition to Prevent Chemical Disasters.

Chapter 7 – Testimony of Dr. M. Sam Mannan, Regents Professor and Director, Mary Kay O'Connor Process Safety Center, Texas A&M University.

In: Ammonium Nitrate Facilities
Editor: Allegra Ward

ISBN: 978-1-63321-716-4

Chapter 1

CHEMICAL SAFETY: ACTIONS NEEDED TO IMPROVE FEDERAL OVERSIGHT OF FACILITIES WITH AMMONIUM NITRATE*

United States Government Accountability Office

WHY GAO DID THIS STUDY

In April 2013, about 30 tons of ammonium nitrate fertilizer detonated during a fire at a facility in West, Texas, killing at least 14 people and damaging nearby schools, homes, and a nursing home. This incident raised concerns about the risks posed by similar facilities across the country. OSHA and EPA play a central role in protecting workers and communities from chemical accidents, and DHS administers a chemical facility security program. GAO was asked to examine oversight of ammonium nitrate facilities in the United States and other countries. This report addresses (1) how many facilities have ammonium nitrate in the United States, (2) how OSHA and EPA regulate and oversee facilities that have ammonium nitrate, and (3) what approaches selected other countries have adopted for regulating and overseeing facilities with ammonium nitrate. GAO analyzed available federal data and data from selected states with high use of ammonium nitrate; reviewed federal laws and regulations; and interviewed government officials,

* This is an edited, reformatted and augmented version of the United States Government Accountability Office publication, GAO-14-274, dated May 2014.

chemical safety experts, and industry representatives in the United States and selected countries.

WHAT GAO RECOMMENDS

GAO is recommending that federal agencies improve data sharing, OSHA and EPA consider revising their related regulations to cover ammonium nitrate, and OSHA conduct outreach to the fertilizer industry and target high risk facilities for inspection. DHS, EPA, and OSHA agreed with GAO's recommendations and suggested technical changes, which GAO incorporated as appropriate.

WHAT GAO FOUND

Federal data provide insight into the number of facilities in the United States with ammonium nitrate but do not provide a complete picture because of reporting exemptions and other data limitations. The Occupational Safety and Health Administration (OSHA) and the Environmental Protection Agency (EPA) do not require facilities to report their ammonium nitrate holdings. The Department of Homeland Security (DHS) requires facilities with certain quantities of ammonium nitrate to report their holdings for security purposes. While the total number of facilities in the United States with ammonium nitrate is unknown, as of August 2013, at least 1,300 facilities in 47 states reported to DHS that they had reportable quantities of ammonium nitrate. Federal law also requires certain facilities to report their ammonium nitrate holdings to state and local authorities for emergency planning purposes, but these data are not routinely shared with federal agencies. According to EPA, states are not required to report these data to federal agencies, and each state determines how to share its data. As part of an Executive Order on Improving Chemical Facility Safety and Security issued in August 2013, federal agencies are exploring options for improving data sharing, but this work is not yet complete.

OSHA and EPA provide limited oversight of facilities that have ammonium nitrate. OSHA's regulations include provisions for the storage of ammonium nitrate, but the agency has done little outreach to increase awareness of these regulations within the fertilizer industry, a primary user. In addition, the regulations have not been significantly revised since 1971 and allow storage of ammonium nitrate in wooden buildings, which could increase the risk of fire and explosion. Other OSHA and EPA chemical safety regulations—which require facilities to complete hazard assessments, use procedures to prevent and respond to accidents, and conduct routine compliance audits—do not apply to ammonium nitrate. Furthermore, although OSHA targets worksites in certain industries for inspection, its inspection programs do not target facilities with ammonium nitrate and, according to OSHA officials, information on these facilities is not available to them to use for targeting the facilities. International chemical safety guidance suggests authorities should provide facilities information on how regulatory requirements can be met and periodically inspect them.

GAO reviewed approaches to overseeing facilities with ammonium nitrate in Canada, France, Germany, and the United Kingdom, selected in part based on recommendations from chemical safety experts. According to foreign officials and government documents, these countries require facilities with specified quantities of ammonium nitrate to assess its risk and develop plans or policies to prevent chemical accidents. For example, Canadian officials said facilities with 22 tons or more of ammonium nitrate are required to complete a risk assessment and an emergency plan. Some countries' storage requirements also restrict the use of wood to store ammonium nitrate. For example, officials told GAO that France restricted the use of wood for storing ammonium nitrate fertilizer after several incidents involving ammonium nitrate fertilizer, and German officials told GAO that certain ammonium nitrate and ammonium nitrate-based preparations must be separated from combustible materials by brick or concrete walls.

ABBREVIATIONS

CFATS	Chemical Facility Anti-Terrorism Standards
DHS	Department of Homeland Security
EPA	Environmental Protection Agency
EPCRA	Emergency Planning and Community Right-to- Know Act of 1986
EU	European Union
OECD	Organisation for Economic Co-operation and Development
OSHA	Occupational Safety and Health Administration
OSH Act	Occupational Safety and Health Act of 1970

PSM Process Safety Management
RMP Risk Management Program

May 19, 2014

The Honorable Barbara Boxer
Chairman
Committee on Environment and Public Works
United States Senate

The Honorable George Miller
Ranking Member
Committee on Education and the Workforce
House of Representatives

The Honorable Robert P. Casey, Jr. Chairman
Subcommittee on Employment and Workplace Safety
Committee on Health, Education, Labor, and Pensions
United States Senate

The Honorable Joe Courtney
Ranking Member
Subcommittee on Workforce Protections
Committee on Education and the Workforce
House of Representatives

On April 17, 2013, about 30 tons of ammonium nitrate fertilizer detonated during a fire at a fertilizer storage and distribution facility in West, Texas, killing at least 14 people and injuring more than 200 others. The explosion severely damaged or destroyed nearly 200 homes, three nearby schools, a nursing home, and an apartment complex.[1] While ammonium nitrate is widely used in agriculture, mining, and other industries, the Texas tragedy underscores the need for great care in its storage and handling. Today, significant quantities of ammonium nitrate fertilizer are stored in facilities across the United States. In 2012, use of ammonium nitrate fertilizer in the United States amounted to 853,093 tons.[2] In 2010, U.S. companies reported producing about 7.5 million tons of ammonium nitrate.[3] The total number and location of facilities in the United States in which ammonium nitrate is stored, however, is not known.

In response to the explosion in West, Texas, President Obama issued an Executive Order on August 1, 2013 designed to improve the safety and security of chemical facilities and reduce the risks that hazardous chemicals pose to workers and communities.[4] The order, which includes a focus on ammonium nitrate, established a federal working group to improve federal coordination with state and local partners; enhance federal agency coordination and information sharing; modernize policies, regulations, and standards; and work with stakeholders to identify best practices.

Several federal agencies are involved in regulating facilities with hazardous chemicals, but the Department of Labor's Occupational Safety and Health Administration (OSHA) and the Environmental Protection Agency (EPA) play central roles in protecting workers and communities from chemical accidents at facilities.[5] In addition, the Department of Homeland Security (DHS) administers a chemical facility security program, the Chemical Facility Anti-Terrorism Standards (CFATS) program, which requires certain chemical facilities to report information to DHS and, in some instances, take additional steps to secure their facilities. You asked us to examine federal oversight of facilities with ammonium nitrate in the United States and approaches used by other countries. For this review we addressed the following questions: (1) How many facilities in the United States have ammonium nitrate? (2) How do OSHA and EPA regulate and oversee facilities that have ammonium nitrate? (3) What approaches have selected other countries adopted for regulating and overseeing facilities with ammonium nitrate?

To answer question 1, we analyzed data from DHS's CFATS program and other sources on the number and types of facilities that reported having ammonium nitrate as of August 2013 and documented the limitations of the data.[6] To assess the reliability of the CFATS data, we reviewed agency documentation, interviewed DHS officials, and performed electronic testing of required data elements. We also requested state data on facilities that reported having ammonium nitrate from four states with high ammonium nitrate fertilizer consumption– Alabama, Missouri, Tennessee, and Texas[7]–and received data from Texas and Alabama.[8] We compared data collected by DHS to other data sources, including chemical inventory data from Alabama and Texas, which were identified as leading users of ammonium nitrate fertilizer, and trade data collected by DHS's Customs and Border Protection agency on U.S. imports and exports of ammonium nitrate.[9] Our primary purpose in comparing CFATS data with data from other sources was to determine whether the CFATS data represent a complete count of facilities with ammonium nitrate. We determined that the CFATS data were sufficiently reliable for purposes of providing the number and type of facilities that reported having ammonium nitrate at levels that met thresholds for reporting under CFATS. As we discuss later in this report, certain limitations of the data did not allow us to determine whether all facilities that should have reported to DHS actually did so.

For question 2, we reviewed relevant federal laws and regulations, focusing on OSHA's and EPA's regulations, including the types of facilities covered by the regulations.[10] We also interviewed federal agency officials regarding their oversight practices.

To describe the approaches selected other countries have adopted for regulating and overseeing facilities with ammonium nitrate, we reviewed approaches used by selected member countries of the European Union (EU) and the Organisation for Economic Co-operation and Development (OECD): Canada, France, Germany, and the United Kingdom. To select these four countries, we considered the extent to which the countries use ammonium nitrate fertilizer, the results of our literature search, and recommendations from our interviews with chemical safety experts. There are key differences between the United States and these other countries, including the size of the country, the size of the agriculture industry, and the amount of ammonium nitrate used. We interviewed government officials from the EU and the countries selected and reviewed documents provided by the officials. We did not conduct an independent legal analysis to verify the information provided about the laws, regulations, or policies of the foreign countries selected for this study. We also interviewed U.S. and international fertilizer industry associations, chemical safety experts, and federal officials to obtain their views on U.S. chemical safety regulations and oversight, and the practices of the selected countries.

We conducted this performance audit from June 2013 to May 2014 in accordance with generally accepted government auditing standards. Those standards require that we plan and perform the audit to obtain sufficient, appropriate evidence to provide a reasonable basis for our findings and conclusions based on our audit objectives. We believe that the evidence obtained provides a reasonable basis for our findings and conclusions based on our audit objectives.

BACKGROUND

Use and Hazards of Ammonium Nitrate

Ammonium nitrate products are manufactured and sold in various forms, depending upon their use. For example, ammonium nitrate fertilizer may be produced and sold in liquid form or as solid granules.[11] According to The Fertilizer Institute, solid ammonium nitrate fertilizer is used heavily by farmers in Alabama, Missouri, Tennessee, and Texas primarily on pastureland, hay, fruit, and vegetable crops.[12] In addition to its agricultural benefits, ammonium nitrate can be mixed with fuel oil or other additives and used by the mining and construction industries as an explosive for blasting.[13]

While ammonium nitrate can increase agricultural productivity, use of this chemical poses a safety and health risk because it can intensify a fire and, under certain circumstances, explode. Ammonium nitrate by itself does not burn, but it increases the risk of fire if it comes in contact with combustible materials. Ammonium nitrate that is stored in a confined space and reaches high temperatures can explode.[14] An explosion is more likely to occur if ammonium nitrate is contaminated by certain materials, such as fuel oil, or if it is stored in large stacks.

Because of ammonium nitrate's potential to facilitate an explosion, facilities storing ammonium nitrate may pose a security threat in part because it can be used to make weapons. Ammonium nitrate fertilizer has been used by domestic and international terrorists to make explosive devices.[15] For example, on April 19, 1995, ammonium nitrate fertilizer—mixed with fuel oil—was used by a domestic terrorist to blow up a federal building in Oklahoma City, Oklahoma. The explosion killed 168 people and injured hundreds more.

Ammonium nitrate has been involved in several major chemical accidents over the past century, including explosions in the United States and Europe. In addition to killing at least 14 people and injuring more than 200 others, the explosion in West, Texas severely damaged or destroyed nearly 200 homes; an apartment complex; and three schools that were, at the time, unoccupied (see figure 1).[16] Prior to that incident, an explosion in 1994 involving ammonium nitrate at a factory in Port Neal, Iowa killed four workers and injured 18 people. In 1947, explosions aboard two ships holding thousands of tons of ammonium nitrate fertilizer killed more than 500 people, injured approximately 3,500, and devastated large areas of industrial and residential buildings in Texas City, Texas. In Europe, accidents involving ammonium nitrate have occurred in Germany, Belgium, and France. A 1921 accident in

Germany and one in Belgium in 1942 caused hundreds of deaths after explosives were used to break up piles of hundreds of tons of ammonium nitrate, resulting in large scale detonations. In France, a ship carrying more than 3,000 tons of ammonium nitrate exploded in 1947, a few months after the Texas City disaster, after pressurized steam was injected into the storage area in an attempt to put out a fire. In 2001, an explosion at a fertilizer plant in Toulouse, France involving between 22 and132 tons of ammonium nitrate resulted in 30 deaths, thousands of injuries requiring hospitalization, and widespread property damage. Past accidents also indicate that smaller quantities of ammonium nitrate can cause substantial damage. For example, in 2003, an explosion of less than 6 tons of ammonium nitrate in a barn in rural France injured 23 people and caused significant property damage.

Federal Agencies' Responsibilities for Promoting Chemical Safety and Security

OSHA and EPA play key roles in protecting the public from the effects of chemical accidents, with EPA focusing on the environment and public health and OSHA focusing on worker safety and health. Under the Occupational Safety and Health Act of 1970 (OSH Act), OSHA is the federal agency responsible for setting and enforcing regulations to protect workers from hazards in the workplace, including exposure to hazardous chemicals.[17] In addition, the Clean Air Act Amendments of 1990 designated roles for both OSHA and EPA with respect to preventing chemical accidents and preparing for the consequences of chemical accidents.[18] In response to requirements in this act, OSHA issued Process Safety Management (PSM) regulations in 1992 to protect workers engaged in processes that involve certain highly hazardous chemicals, and EPA issued Risk Management Program (RMP) regulations in 1996 to require facilities handling particular chemicals to plan how to prevent and address chemical accidents.[19] The PSM and RMP regulations each apply to processes involving a specified list of chemicals above threshold quantities, and require covered facilities to take certain steps to prevent and prepare for chemical accidents. However, neither OSHA's PSM regulations nor EPA's RMP regulations cover ammonium nitrate.

Source: Chemical Safety Board.

Figure 1. Photographs of Damage from the Explosion in West, Texas in April 2013.

The Emergency Planning and Community Right-to-Know Act of 1986 (EPCRA) establishes authorities for emergency planning and preparedness and emergency release notification reporting, among other things.[20] Under section 312 of EPCRA and EPA regulations, facilities with certain hazardous chemicals in amounts at or above threshold levels— including ammonium nitrate in some circumstances—are required to annually submit chemical inventory forms to state and local authorities to help emergency response officials prepare for and respond to chemical incidents.[21]

For purposes of enhancing chemical facility security, the Department of Homeland Security (DHS) Chemical Facility Anti-Terrorism Standards (CFATS) program requires facilities possessing certain chemicals at or above threshold quantities–including some types of ammonium nitrate–to submit reports to DHS with information about the facility and the regulated chemicals present on site.[22] Among other things, DHS collects information on the quantities of certain hazardous chemicals held at facilities, the location of the facilities, and their industry codes.[23] DHS set different threshold quantities for reporting based on the type of ammonium nitrate and the type of security risk presented (see table 1).

Table 1. Thresholds for Reporting Ammonium Nitrate under the CFATS Program

Type of Ammonium Nitrate	Reporting Threshold
Ammonium nitrate with more than 0.2 percent combustible substances, including any organic substance calculated as carbon, to the exclusion of any other added substance[a]	400 pounds or more for a theft risk (if in transportation packaging) [c] and 5,000 pounds or more for a release risk
Solid ammonium nitrate with a nitrogen concentration of 23 percent or greater, and, if in a mixture, a minimum ammonium nitrate concentration of 33 percent or greater [b]	2,000 pounds or more for a theft risk (if in transportation packaging)

Source: CFATS regulations,. 6 C.F.R. pt. 27 and app. A.

[a] According to DHS, this type of ammonium nitrate is more commonly used as an explosive and is regulated by the Department of Transportation as a Division 1.1 explosive. Division 1.1 consists of explosives that have a mass explosion hazard. A mass explosion is one which affects almost the entire load instantaneously. 49 C.F.R. § 173.50(b)(1).

[b] According to DHS, this type of ammonium nitrate is more commonly used by the agricultural community as a fertilizer; however, it may be compounded with other ingredients to create an explosive.

[c] DHS's CFATS regulations provide that in calculating whether a facility possesses a threshold amount of a chemical that poses a theft or diversion risk, the facility shall only include those chemicals that are in transportation packaging as defined by Department of Transportation regulations. 6 C.F.R. § 27.203(c).

Not all facilities with ammonium nitrate, however, are required to file CFATS reports with DHS. First, facilities are only required to report if they are holding amounts equal to or greater than threshold quantities of specific types of ammonium nitrate. Also, DHS does not require certain agricultural producers to report their chemical holdings to DHS.[24] In addition, DHS's reporting threshold for ammonium nitrate fertilizer only applies to quantities held in transportable containers such as cylinders, bulk bags, bottles (inside or outside of boxes), cargo tanks, and tank cars.[25] Finally, there are several statutory exemptions to CFATS requirements. Specifically, CFATS does not apply to public water systems or treatment works, any facility that is owned or operated by the Department of Defense or the Department of Energy, facilities regulated by the Nuclear Regulatory Commission, or facilities covered by the Maritime Transportation Security Act of 2002 administered by the Coast Guard.[26]

Other federal agencies regulate different aspects of the use of hazardous chemicals. For example, the Department of Transportation regulates the transport of hazardous materials, the Coast Guard inspects containers of hazardous materials at ports and waterways, and the Bureau of Alcohol, Tobacco, Firearms and Explosives (ATF) in the Department of Justice regulates the manufacture, distribution, and storage of explosive materials, including blasting agents and other explosive materials containing ammonium nitrate.[27]

State and Local Government Responsibilities for Promoting Chemical Safety

State and local government agencies are also involved in regulating hazardous chemical facilities under federal laws and their own state or local laws. Federal laws may authorize or assign state and local governments certain roles and responsibilities for overseeing chemical facilities. For example, as permitted by the OSH Act, OSHA has approved state plans that authorize about half the states to operate their own occupational safety and health programs.[28] As a result, private sector workplaces in 21 states and Puerto Rico are regulated and inspected by state occupational safety and health agencies rather than OSHA.[29]

Similarly, EPA has delegated its authority to implement and enforce the Risk Management Program to nine states and five counties.[30] As previously mentioned, both state and local governments play a role in implementing

EPCRA, which requires covered facilities to report basic information about their hazardous chemical inventories to certain state and local authorities, including estimates of the amounts of chemicals present at facilities.

In addition, state and local governments may establish and enforce their own laws, regulations, or ordinances to protect the public from chemical accidents.

For example, state and local governments may adopt and enforce fire codes or zoning laws that specify how far chemical facilities must be located from residential areas.

Executive Order on Improving Chemical Facility Safety and Security

The Executive Order issued on August 1, 2013 established a Chemical Facility Safety and Security Working Group co-chaired by the Secretary of Homeland Security, the Administrator of EPA, and the Secretary of Labor. The Executive Order includes directives for the working group to: improve operational coordination with state and local partners; enhance federal agency coordination and information sharing; modernize policies, regulations, and standards; and work with stakeholders to identify best practices. The order includes tasks focused specifically on ammonium nitrate.[31] Specifically, it directs the Secretaries of Homeland Security, Labor, and Agriculture to develop a list of potential regulatory and legislative proposals to improve the safe and secure storage, handling, and sale of ammonium nitrate. In addition, the Department of Labor and EPA are directed to review the chemical hazards covered by the RMP and PSM regulations and determine whether they should be expanded to address additional hazards.

OECD's Guidance on Chemical Safety

The Organisation for Economic Co-operation and Development (OECD), an intergovernmental organization with 34 member countries, issued guidance in 2003 on the prevention of, preparedness for, and response to chemical accidents.[32]

This publication was developed with other international organizations active in the area of chemical accident safety, such as the World Health Organization.

The document—OECD Guiding Principles for Chemical Accident Prevention, Preparedness and Response—includes detailed guidance for industry, public authorities, and the public on how they can help prevent chemical accidents and better respond when accidents occur.

OVER 1,300 FACILITIES IN 47 STATES REPORTED HAVING AMMONIUM NITRATE, BUT DATA LIMITATIONS PREVENT OBTAINING A COMPLETE COUNT OF FACILITIES

The total number of facilities in the United States with ammonium nitrate is not known because of the different reporting criteria used by different government agencies, reporting exemptions, and other data limitations. While the total number is unknown, over 1,300 facilities reported having ammonium nitrate to DHS.

DHS's data, however, do not include all facilities that work with ammonium nitrate, in part because some facilities, such as farms, currently do not have to report to DHS and, according to DHS officials, other facilities that are required to report may fail to do so.

DHS Data List Over 1,300 Facilities in 47 States with Ammonium Nitrate

As of August 2013, 1,345 facilities located in 47 states reported to DHS under CFATS that they had ammonium nitrate. The facilities that reported to DHS as having reportable quantities of ammonium nitrate were most often engaged in supplying and supporting the agriculture and mining industries. Many of these facilities were concentrated in the South. About half of these facilities were located in six states: Alabama, Georgia, Kentucky, Missouri, Tennessee, and Texas.

Table 2 shows the number of facilities that reported to DHS that they had ammonium nitrate and the number of states in which they were located.

Table 2. Number of Facilities that Reported Having Ammonium Nitrate to the Department of Homeland Security (DHS) and the Number of States in Which They Were Located, August 2013

Type of Ammonium Nitrate	Number of Facilities	Number of States
Ammonium nitrate with more than 0.2 percent combustible substances, including any organic substance calculated as carbon, to the exclusion of any other added substance[a]	230	42
Solid ammonium nitrate with a nitrogen concentration of 23 percent or greater, and, if in a mixture, a minimum ammonium nitrate concentration of 33 percent or greater [b]	941	45
Reported having both types of ammonium nitrate	174	40
Total number of facilities that reported having ammonium nitrate	**1,345**	**47**

Source: GAO analysis of DHS data.

[a] According to DHS, this type of ammonium nitrate is more commonly used as an explosive and is regulated by the Department of Transportation as a Division 1.1 explosive. The threshold quantity for reporting this type of ammonium nitrate is 400 pounds or more for a theft risk (if in transportable packaging) and 5,000 pounds or more for a release risk.

[b] According to DHS, this type of ammonium nitrate is more commonly used by the agricultural community as a fertilizer; however, it may be compounded with other ingredients to create an explosive. The threshold quantity for reporting this type of ammonium nitrate for a theft risk (if in transportation packaging) is 2,000 pounds or more.

State Data and Federal Trade Data Suggest That the Total Number of Facilities with Ammonium Nitrate is Greater Than Those That Report to DHS

Our review of additional state data, including EPRCA data,[33] from Texas and Alabama, which have different reporting criteria than CFATS, indicated that there are more facilities with ammonium nitrate than those that report to DHS.[34] We compared the data they provided to the data on facilities that reported to DHS under CFATS. In these two states, we found that the data from each of the sources provided to us differed and that no single count of such facilities, whether from the state or DHS, represented a comprehensive picture of facilities with ammonium nitrate.

For Texas, we reviewed three sources of data on facilities that have ammonium nitrate: (1) EPCRA data from the Texas Department of State Health Services; (2) a list of facilities that registered with the Office of the Texas State Chemist as having plans to produce, store, or sell ammonium nitrate; [35] and (3) DHS's CFATS data. We compared data from all three of these sources and found 189 facilities that reported having ammonium nitrate (see figure 2). Of these 189 facilities, 52 filed CFATS reports with DHS. Data were not readily available to determine whether the remaining facilities were required to file CFATS reports. DHS officials told us the agency has begun an effort to obtain lists of chemical facilities the states have compiled and compare them with its CFATS data to identify facilities that should have filed CFATS reports but did not. This effort is still under way. As shown in figure 2, 17 of the 189 facilities in Texas were listed in all three data sources.

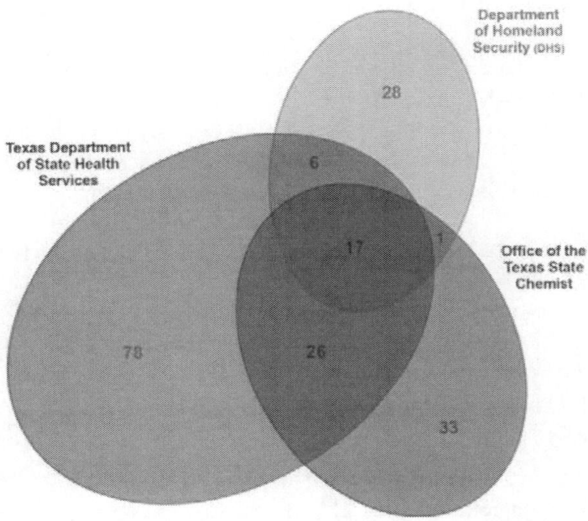

Source: GAO analysis of U.S. Department of Homeland Security and state agency data.

Note: This figure includes data reported to the Texas Department of State Health Services under EPCRA section 312 as of December 2012, data collected by the Office of the Texas State Chemist under state law as of November 2013, and data reported to DHS under the CFATS program as of August 2013. Each of these programs has different reporting criteria, therefore, facilities required to report under one program may not be required to report under another program.

Figure 2. Number of Facilities in Texas that Reported to State Agencies and DHS That They Had Ammonium Nitrate.

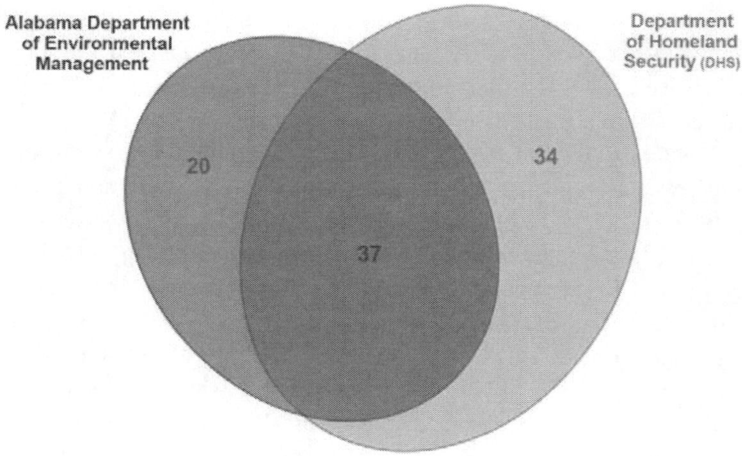

Source: GAO analysis of U.S. Department of Homeland Security and state agency data.

Note: This figure includes data reported to the Alabama Department of Environmental Management under EPCRA section 312 as of December 2012 and data reported to DHS under the CFATS program as of August 2013. Each of these programs has different reporting criteria, therefore, facilities required to report under one program may not be required to report under another program.

Figure 3. Number of Facilities in Alabama that Reported to the State and DHS That They Had Ammonium Nitrate.

For Alabama, we reviewed data from two sources on facilities that reported having ammonium nitrate: (1) EPCRA data from Alabama's Department of Environmental Management, and (2) DHS's CFATS data.[36] From these two sources, we found 91 facilities that reported having ammonium nitrate— 57 that filed EPCRA reports with the state Department of Environmental Management and 71 that filed CFATS reports with DHS. Thirty-seven of the facilities filed reports with both the state and DHS. (See figure 3.)

Our analysis of federal trade data collected by DHS's Customs and Border Protection agency also suggests that a greater number of facilities have ammonium nitrate than those that reported to DHS under the CFATS program.[37] Using the data from the Customs and Border Protection agency, we identified 205 facilities that imported ammonium nitrate products and 81 facilities that exported ammonium nitrate products in fiscal year 2013.[38] The majority of these facilities reported importing or exporting mixtures of ammonium nitrate and calcium carbonate or mixtures of urea and ammonium nitrate. Eight of these facilities filed CFATS reports with DHS. Moreover, we found about 100 facilities that imported or exported a form of ammonium nitrate that may be subject to DHS's CFATS requirements for reporting quantities over 2,000 pounds but did not file a report.[39] These facilities, however, may not be required to file CFATS reports. For example, they may meet one of the statutory exemptions, or the composition of their ammonium nitrate (or their ammonium nitrate mixture) may not trigger the reporting requirements. Data were not readily available to determine whether they met all of DHS's reporting requirements for the CFATS program. In addition, according to DHS officials, other data limitations could explain some of the differences between the CFATS data and the federal trade data. For example, facilities may submit reports to the different agencies using different names and addresses. According to DHS, different people in the facility may prepare the different reports; the facility may define the perimeters of each site differently; or the corporate structure or nomenclature may have changed from the time one report was submitted to the next reporting period.

Differences in Reporting Criteria, Including Exemptions, and Facilities' Failure to File Required Reports Also Prevent Full Identification of Facilities with Ammonium Nitrate

The total number of facilities with ammonium nitrate is also difficult to determine because of the variation in reporting criteria, including exemptions for some facilities from reporting to either their state or to DHS. For example, farmers could be exempt from reporting under both EPCRA and CFATS because EPCRA's reporting requirements do not apply to substances used in routine agricultural operations and DHS does not currently require certain agricultural producers to report their chemical holdings to DHS.[40] In addition, DHS's reporting threshold for ammonium nitrate fertilizer only applies to quantities held in transportable containers such as cylinders, bulk bags, bottles (inside or outside of boxes), cargo tanks, and tank cars.[41] Also, EPCRA does not require retailers to report fertilizer held for sale to the ultimate customer. However, an August 2013 chemical advisory on ammonium nitrate issued jointly by EPA, OSHA, and ATF clarified that EPCRA requires fertilizer distributors to report ammonium nitrate that is blended or mixed with other chemicals on site. In addition, some facilities may not report to DHS or their state because they have amounts of ammonium nitrate that are below the applicable reporting thresholds.

Some facilities may not be included in either DHS's or states' data because they fail to submit their required reports, but the magnitude of underreporting is not known. DHS officials acknowledged that some facilities fail to file the required forms. The facility in West, Texas had not filed a CFATS report to DHS but, in 2012, this facility filed an EPRCA form with the state, reporting that it had 270 tons of ammonium nitrate. According to DHS officials, the agency does not know with certainty whether the West, Texas facility should have reported its ammonium nitrate to DHS because the agency did not visit the facility after the explosion and it does not know the manner in which the facility held its ammonium nitrate prior to the explosion. Following the explosion at the facility in West, Texas, DHS obtained data from the state of Texas and compared the state data to the facilities that reported to DHS. As a result of this data matching effort, DHS sent out 106 letters to other potentially noncompliant facilities in Texas. According to DHS, many of the Texas facilities that received the letter said they do not actually possess ammonium nitrate or do not meet the criteria to require reporting under CFATS. DHS has used EPA's Risk Management Program (RMP) database to try and identify such facilities holding other chemicals, but it cannot use the RMP database to identify all facilities with ammonium nitrate because ammonium nitrate is not covered by EPA's RMP regulations. In addition, DHS officials told us the agency is in the process of comparing its list of facilities that reported to DHS under the CFATS program to ATF's list of facilities that have federal explosives licenses and permits to identify potentially noncompliant facilities, but this effort had not been completed at the time of our review.

OSHA Lacks Access to Data on Facilities That Have Ammonium Nitrate

OSHA has limited access to data collected by other agencies to use in identifying facilities with ammonium nitrate. DHS does not currently share its CFATS data with OSHA, although DHS officials told us they were not aware of anything prohibiting DHS from doing so. While EPA shares data from its RMP with OSHA on a quarterly basis, the data do not include information on ammonium nitrate because ammonium nitrate is not covered by EPA's RMP regulations. As previously discussed, under section 312 of EPCRA, facilities are required to annually report information to state and local authorities on the types and quantities of certain hazardous chemicals present at their facilities, which may include ammonium nitrate. Facilities that possess reportable quantities of ammonium nitrate submit this information electronically or on paper forms, and the state and local entities maintain copies of these forms. However, according to agency officials, the EPCRA data are not shared directly with federal agencies, including OSHA, EPA, or DHS (see figure 4). EPA officials, however, noted that EPCRA is primarily intended to provide information to state and local officials, not to other federal agencies. Any person may submit written requests to the designated state or local authority for information on individual facilities that may have ammonium nitrate, but lists of all facilities in a state that have submitted these data, including those that reported having ammonium nitrate, are not publicly available.[42] In certain states we contacted, officials indicated that data on individual facilities could be requested from the state, but the requester would have to request data on specific facilities to obtain information on the chemicals they hold. OSHA officials cited the lack of access to data on facilities with ammonium nitrate as a reason they would have difficulty designing an inspection program to target such facilities.

The University of Texas at Dallas has a database (called E-Plan) that contains EPCRA data from over half of the states for the 2012 reporting year, but federal agencies have made limited use of it. University staff originally developed the E-Plan database in 2000 with funding from EPA to facilitate EPCRA reporting and provide first responders rapid access to information on chemical facilities in emergency situations. In many local areas, first responders and emergency services personnel can use the E- Plan data when they prepare for and respond to emergencies such as fires. According to E-Plan administrators, OSHA staff helped develop the database, but currently OSHA does not use E-Plan. EPA staff told us that some EPA regional offices have used the E-Plan database to assess compliance with the agency's RMP reporting requirements. DHS officials told us the agency does not use E-Plan data to assess compliance with CFATS requirements. DHS officials also explained that, while the database could contain useful information, it is incomplete. Some states do not submit data to E-Plan at all, and other states' data are incomplete. In addition, participation in E-Plan is voluntary and, even among those states that participate, some states do not choose to allow their E-Plan data to be shared with federal agencies.

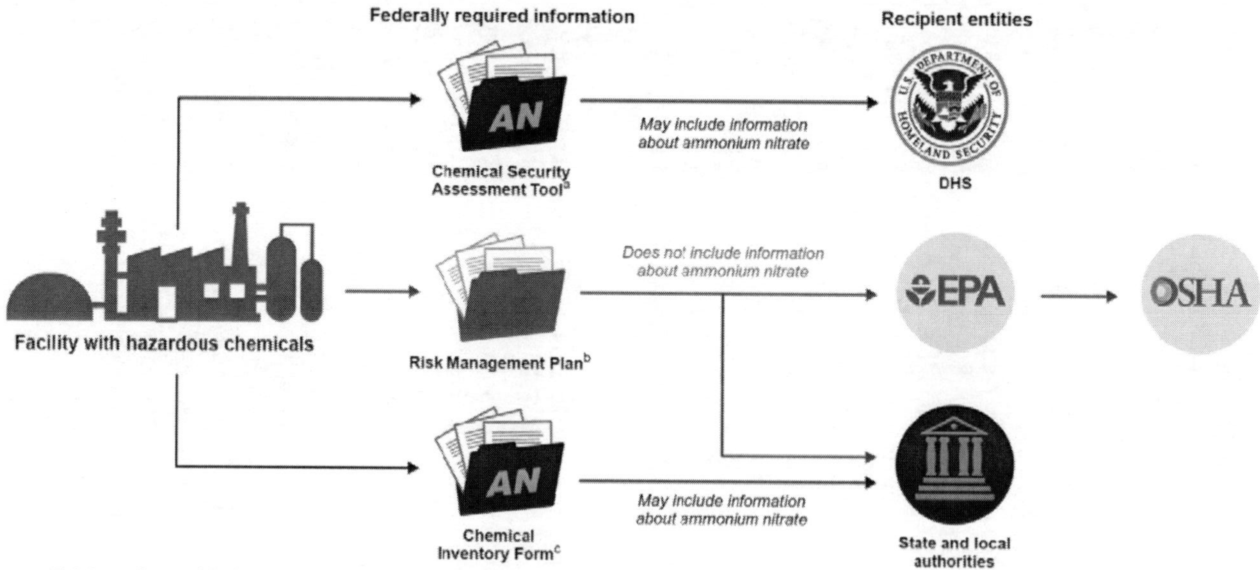

Source: GAO review of federal regulations and interviews with federal officials.

[a] DHS's Chemical Security Assessment Tool is used for submitting reports under DHS's CFATS program.

[b] A risk management plan is required under EPA's RMP regulations.

[c] The Chemical Inventory Form is used for submitting reports under section 312 of EPCRA. EPA publishes model forms; however, alternative formats are permitted provided they comply with EPCRA and EPA's regulations.

Note: Facilities are only required to report if they meet the reporting requirements for each program.

Figure 4. Entities Receiving Federally Required Hazardous Chemical Reports from Facilities.

Ongoing Efforts to Improve Data Sharing on Chemical Facilities

The Chemical Facility Safety and Security Working Group established by the August 2013 Executive Order has begun its efforts to develop proposals for improving information sharing, but this work has not been completed. The working group has held listening sessions throughout the country seeking input from interested parties on options for making improvements in chemical safety and security. It also has launched a pilot program in New York and New Jersey aimed at improving access to data on chemical facilities for federal, state, local, and tribal governments. In addition, the working group is evaluating how federal agencies can work with states to enhance the states' roles as information sharing organizations, including options for sharing RMP, CFATS, and EPCRA data. Finally, it is exploring ways for federal and state agencies to share information and exchange data to, among other things, identify chemical facilities that are not in compliance with safety and security requirements. For example, DHS and EPA are comparing their CFATS and RMP data to determine if the CFATS data include facilities that should also have reported under the RMP. As a result, EPA has begun sending notification letters to facilities requesting information to help determine if the facility is subject to RMP requirements. Because the RMP regulations do not currently cover ammonium nitrate, however, this strategy would not be useful for identifying facilities that have ammonium nitrate. The federal working group is also sharing information to, among other things, identify whether additional facilities have failed to report under CFATS and is exploring whether EPA software offered to states to facilitate EPCRA reporting could also provide a vehicle to enhance access to the reports while meeting security objectives.

OSHA Has Not Focused Its Enforcement Efforts on Ammonium Nitrate and EPA Has Not Regulated It as a Hazardous Material

OSHA has regulations for the storage of ammonium nitrate, but the agency has not focused its enforcement resources on the use of ammonium nitrate by the fertilizer industry, which is a primary user. EPA, on the other hand, has regulations requiring risk management planning by facilities that have certain hazardous chemicals, but these regulations do not apply to ammonium nitrate.[43]

OSHA's Regulations for Explosives and Blasting Agents List Substantive Requirements for the Storage of Ammonium Nitrate

OSHA's Explosives and Blasting Agents regulations—issued in 1971—include provisions for the storage of both explosives grade and fertilizer grade ammonium nitrate in quantities of 1,000 pounds or more.[44] OSHA based these regulations on two 1970 consensus standards developed by the National Fire Protection Association (NFPA).[45] Few significant changes have been made to these regulations since they were issued, although the National Fire Protection Association periodically reviews and updates its standards. OSHA's regulations include requirements that could reduce the fire and explosion hazards associated with ammonium nitrate, such as required fire protection measures, limits on stack size, and requirements related to separating ammonium nitrate from combustible and other contaminating materials. However, the regulations do not categorically prohibit employers from storing ammonium nitrate in wooden bins and buildings.[46] In addition, if the facilities were in existence at the time the regulations were issued in 1971, OSHA's regulations allow the use of storage buildings not in strict conformity with the regulations if such use does not constitute a hazard to life.[47] Some of the provisions of OSHA's ammonium nitrate storage regulations are described in table 3.

Recently, OSHA, EPA, and ATF jointly issued a chemical advisory that recommends that facilities store ammonium nitrate in non-combustible buildings.[48] Similarly, following the explosion in West, Texas, the National Fire Protection Association is considering changes to its ammonium nitrate storage provisions, which are part of its hazardous materials consensus standard, including restricting the use of wood to store ammonium nitrate.

In addition to storage requirements, OSHA's Hazard Communication regulations require that employers whose workers are exposed to hazardous chemicals, including ammonium nitrate, inform their workers of the dangers and train them to handle the materials appropriately. Employers are required to use labels, training, and safety data sheets to inform workers of chemical hazards in the workplace.[49] Safety data sheets are written documents with details on the hazards associated with each chemical, measures workers can take to protect themselves, actions workers should take in case of an emergency, and safety precautions for handling and storing the chemical.

Table 3. Ammonium Nitrate Storage Topics Addressed in Selected Provisions of OSHA's Explosives and Blasting Agents Regulations

Topic	Summary of Selected Provisions	Citation(s)
Who must comply	All persons storing, having, or keeping ammonium nitrate, and the owner or lessee of any building, premises, or structure in which ammonium nitrate is stored in quantities of 1,000 pounds or more. Applies to the storage of [solid] ammonium nitrate, including fertilizer grade, dynamite grade, nitrous oxide grade, technical grade, and other mixtures containing 60 percent or more ammonium nitrate by weight, but does not apply to blasting agents. Certain additional exceptions apply.	29 C.F.R. § 1910.109(i)(2)(i). 29 C.F.R. § 1910.109 (i)(1)(i)(a). 29 C.F.R. § 1910.109 (i)(1)(i)(b)-(c).
Storage building construction	The wall on the exposed side of a storage building within 50 feet of a combustible building, forest, piles of combustible materials and similar exposure hazards shall be of fire-resistive construction. In lieu of the fire-resistive wall, other suitable means of exposure protection such as a free standing wall may be used. All flooring in storage and handling areas shall be of noncombustible material or protected against impregnation by ammonium nitrate and shall be without open drains, traps, tunnels, pits, or pockets into which any molten ammonium nitrate could flow and be confined in the event of fire. The continued use of an existing storage building or structure not in strict conformity with [these provisions] may be approved in cases where such continued use will not constitute a hazard to life.	29 C.F.R. § 1910.109 (i)(2)(iii)(c). 29 C.F.R. § 1910.109 (i)(2)(iii)(d). 29 C.F.R. § 1910.109 (i)(2)(iii)(e).
Size of piles and separation distances, when stored in bags, drums or other containers	Minimum distance from walls (bags): 30 inches. Maximum pile height and width: 20 feet. Maximum pile length: 50 feet. Where the building is of noncombustible constriction or is protected by automatic sprinklers the length of the piles is not limited.	29 C.F.R. § 1910.109 (i)(3)(ii)(b). 29 C.F.R. § 1910.109 (i)(3)(ii)(c).

Table 3. (Continued)

Topic	Summary of Selected Provisions	Citation(s)
	Minimum distance from the roof: 36 inches. Aisles shall be provided to separate piles by a clear space of at least 3 feet. At least one service or main aisle in the storage area shall be not less than 4 feet wide.	29 C.F.R. § 1910.109 (i)(3)(ii)(d).
Storage bin construction for bulk ammonium nitrate	Due to the corrosive and reactive properties of ammonium nitrate, and to avoid contamination, galvanized iron, copper, lead, and zinc shall not be used in a bin construction unless suitably protected. Aluminum bins and wooden bins protected against impregnation by ammonium nitrate are permissible. The partitions dividing the ammonium nitrate storage from other products which would contaminate the ammonium nitrate shall be of tight construction.	29 C.F.R. § 1910.109 (i)(4)(ii)(b).
Separation from combustible and other contaminating materials	Ammonium nitrate shall be in a separate building or shall be separated by approved type firewalls of not less than 1 hour fire-resistance rating from storage of organic chemicals, acids, or other corrosive materials, materials that may require blasting during processing or handling, compressed flammable gases, flammable and combustible materials or other contaminating substances.	29 C.F.R. § 1910.109 (i)(5)(i)(a).
Fire protection	Not more than 2,500 tons of bagged ammonium nitrate shall be stored in a building or structure not equipped with an automatic sprinkler system. Suitable fire control devices such as small hose or portable fire extinguishers shall be provided throughout the warehouse and in the loading and unloading areas. Water supplies and fire hydrants shall be available in accordance with recognized good practices.	29 C.F.R. § 1910.109 (i)(7)(i). 29 C.F.R. § 1910.109(i) (7)(ii)(a). 29 C.F.R. § 1910.109 (i)(7)(ii)(b).

Source: Ammonium nitrate storage provisions of OSHA's regulations, 29 C.F.R. § 1910.109(i).

Note: This table is not intended to be comprehensive; additional requirements or exceptions may apply to each topic that are not described here. States with their own OSHA-approved occupational safety and health program must have state standards that are at least as effective as OSHA's.

OSHA Has Conducted Little Outreach to the Fertilizer Industry to Increase Awareness of Its Ammonium Nitrate Storage Regulations

Until the explosion in West, Texas, OSHA had not reached out to the fertilizer industry to inform its members of OSHA's requirements for the storage of ammonium nitrate fertilizer. An OSHA official told us the agency has not traditionally informed the fertilizer industry about these regulations. However, another OSHA official said agency officials met with industry representatives after the explosion at the facility in West, Texas and, based on that meeting, concluded that the fertilizer industry is "well aware" of the agency's storage regulations. OECD's Guiding Principles for Chemical Accident Prevention, Preparedness, and Response recommend that public authorities provide clear, easy-to- understand guidance to facilities on how regulatory objectives and requirements can be met.

OSHA recently published information about how the agency's Explosives and Blasting Agents regulations apply to ammonium nitrate fertilizer. The agency provides employers with training, technical assistance, and information through its website on a variety of safety and health topics.[50] Recently, OSHA updated its website to refer to its storage regulations for ammonium nitrate fertilizer. The August 2013 chemical advisory contains information on OSHA's ammonium nitrate storage regulations, stating that OSHA's Explosives and Blasting Agents regulations contain requirements for the storage of all grades of ammonium nitrate, including fertilizer grade ammonium nitrate. In addition, in February 2014, OSHA announced that the agency is working with the fertilizer industry to remind employers of the importance of safely storing and handling ammonium nitrate. OSHA published a letter on its website that provides employers with legal requirements and best practice recommendations for safely storing and handling ammonium nitrate. In the letter, OSHA states that the agency will enforce the requirements of 29 C.F.R. § 1910.109(i) for storage of ammonium nitrate, including at facilities in non-explosives industries. According to the announcement, fertilizer industry associations will share the letter with facilities across the country.

Fertilizer industry representatives we interviewed said that, prior to the explosion in West, Texas, they did not know that OSHA's ammonium nitrate storage regulations applied to the fertilizer industry, and they suggested that OSHA reach out to the fertilizer industry to help prevent another incident. Industry representatives explained that their understanding was based on a proposed rule published by OSHA in the Federal Register on April 13, 2007, which proposed revisions to the Explosives and Blasting Agents regulation.[51] In that notice, OSHA proposed a change to the ammonium nitrate storage requirements "to clarify that OSHA intends the requirements to apply to ammonium nitrate that will be used in the manufacture of explosives." Although this proposed rule was never finalized, the industry representatives told us they relied on this statement to mean OSHA did not intend the storage requirements to apply to ammonium nitrate fertilizer.

In addition, we reviewed the safety data sheets developed by four U.S. producers of solid ammonium nitrate fertilizer and found that only one company's sheet listed OSHA's Explosives and Blasting Agents regulations as applicable to the storage and handling of ammonium nitrate fertilizer.[52] An industry representative who assists agricultural retailers with regulatory compliance said he reviewed the regulatory information sections in his clients' safety data sheets for ammonium nitrate fertilizer and none of them referred to OSHA's Explosives and Blasting Agents regulations. A representative from one national fertilizer industry association said it would be helpful if OSHA took additional steps to explain its interpretation of the applicable requirements and reach out to the fertilizer industry so that affected companies are better informed. A representative from another national agricultural industry group suggested that OSHA develop and disseminate a compliance assistance tool or checklist to ensure that facilities are aware of and in compliance with the applicable regulations.

The fertilizer industry is developing a voluntary program called Responsible Ag to promote compliance with federal regulations among fertilizer facilities. Officials from the Fertilizer Institute and the Agricultural Retailers Association told us they plan to consolidate federal regulatory requirements for fertilizer retail facilities into one comprehensive checklist and provide third party audits to retailers based on a checklist they have developed. In addition, officials with the Asmark Institute, a nonprofit resource center for agricultural retailers in the United States, said they developed their own compliance assessment tool for agricultural retailers. The Fertilizer Institute and the Agricultural Retailers Association selected the As mark Institute to develop a database that will include information on audit reports and scores from the third party audits. This initiative will be modeled after a voluntary audit program in Minnesota for agricultural retailers to help them improve compliance with federal and state regulations. According to OSHA officials, OSHA has not been involved in the development of this industry initiative.

OSHA Has No Program for Targeted Inspections of Facilities with Ammonium Nitrate

Although OSHA has a national enforcement program that targets certain chemical facilities for inspection, this program does not systematically cover facilities with ammonium nitrate. OECD chemical safety guidance suggests public authorities periodically inspect the safety performance of hazardous facilities.

OSHA conducts inspections of worksites, as authorized under the OSH Act.[53] As part of its enforcement efforts, OSHA randomly selects facilities for inspection as part of a national emphasis program for chemical facilities it initiated in 2011.

However, these inspections are for facilities and chemicals covered under its Process Safety Management (PSM) regulations, which do not include ammonium nitrate. According to OSHA officials, facilities that blend and store ammonium nitrate fertilizer fall outside the scope of this national emphasis program. When we asked whether OSHA might expand its national emphasis program to focus on ammonium nitrate fertilizer facilities, officials said that the agency is not planning on targeting these facilities, in part because OSHA has no means of identifying them.[54]

In addition, OSHA is not likely to target facilities with ammonium nitrate for inspection because of its limited resources, and because these facilities often do not meet OSHA's current inspection priorities. OSHA conducts inspections with its own personnel and the number of inspections OSHA and the states can perform each year is limited by the size of their inspection workforce.

According to OSHA officials, OSHA and the states have about 2,200 inspectors who inspected about 1 percent of the 8 million covered employers in fiscal year 2012. Among OSHA's highest priorities for inspecting worksites are responding to major accidents and employee complaints. In fiscal year 2012, OSHA reported that 44 percent of the agency's inspections were unplanned inspections, which include inspections initiated in response to an accident or complaint.

OSHA also targets certain industries for planned inspections that have high rates of workplace injury and illness. For example, OSHA reported that 55 percent of OSHA's planned inspections in fiscal year 2012 were inspections of worksites in the construction industry.

OSHA has rarely issued citations for violations of its ammonium nitrate storage regulations at fertilizer facilities. OSHA officials told us a citation for a violation of the agency's ammonium nitrate storage regulations was issued as the result of an inspection of a fertilizer facility only once before the explosion in West, Texas.[55] In that case, OSHA inspected a Florida- based fertilizer manufacturer in 1997 in response to a complaint, and cited the company for 30 violations, one of which was a violation of its ammonium nitrate storage requirements.

In addition, according to OSHA officials, within the last 5 years, none of the 21 states that operate their own safety and health programs have cited any employers for improper storage or handling of ammonium nitrate.

OSHA Is Prohibited from Inspecting Some Facilities with Ammonium Nitrate That Have 10 or Fewer Employees

Under a provision regularly included in the annual appropriations act, OSHA is prohibited from conducting planned safety inspections of small employers—those with 10 or fewer employees—in certain low hazard industries, as determined by their injury and illness rates.[56] Although the number of facilities exempted from OSHA inspections under this provision is unclear, we found that, of the facilities that reported having ammonium nitrate to DHS as of August 2013, 60 facilities—about 4 percent of the 1,345 facilities that reported to DHS— reported having 10 or fewer employees and had an industry code with a lower than the average workplace injury and illness rate (see table 4).[57]

As a result, according to OSHA officials, this provision could have hindered the agency's enforcement of its ammonium nitrate storage regulations at these facilities.

OSHA's fiscal year 2015 budget request asks Congress to consider amending OSHA's appropriation language to allow the agency to perform targeted inspections of small establishments that have the potential for catastrophic incidents, such as those with processes covered by OSHA's PSM or EPA's RMP regulations. In the budget request, OSHA states that the current appropriations language limits the agency's ability to conduct inspections, and neither the number of workers in a company nor low injury and illness rates is predictive of the potential for catastrophic accidents that can damage whole communities.

Table 4. Number of Facilities Reporting More Than Threshold Amounts of Ammonium Nitrate to DHS That May be Exempt from Planned OSHA Safety Inspections Based on Industry Classification and Number of Employees

NAICS industry code	NAICS industry code description	Number of facilities with this code that are potentially exempt from programmed inspection based on reporting 10 or fewer employees
111140	Wheat Farming	2
111199	All Other Grain Farming	1
115112	Soil Preparation, Planting, and Cultivating	21
213113	Support Activities for Coal Mining	2
213115	Support Activities for Nonmetallic Minerals (except Fuels)	7
238910	Site Preparation Contractors	4
325120	Industrial Gas Manufacturing	4
325311	Nitrogenous Fertilizer Manufacturing	1
423820	Farm and Garden Machinery and Equipment Merchant Wholesalers	2
424690	Other Chemical and Allied Products Merchant Wholesalers	14
482112	Short Line Railroads	1
541380	Testing Laboratories	1
Total		**60**

Source: GAO review of DHS data reported as of August 2013 and industry codes listed in OSHA's directive regarding the Enforcement Exemptions and Limitations under the Appropriations Act, CPL 02-00-051, changes to Appendix A (effective February 22, 2013).

Note: Facilities are generally required to report to DHS if they have 400 pounds or more of ammonium nitrate meeting the definition of ammonium nitrate regulated by the Department of Transportation as a Division 1.1 Explosive or 2,000 pounds or more of ammonium nitrate commonly used as a fertilizer contained in transportation packaging. Facilities are listed in this table if they (1) reported a North American Industry Classification System (NAICS) code to DHS that had a workplace injury and illness rate below the national average as of 2011, and (2) reported having 10 or fewer employees.

Other OSHA and EPA Chemical Safety Regulations Do Not Apply to Facilities with Ammonium Nitrate

OSHA's PSM regulations for chemical safety do not cover ammonium nitrate.[58] In response to a requirement in the Clean Air Act Amendments of 1990, OSHA issued its PSM regulations in 1992 to help prevent accidents involving highly hazardous chemicals, including toxic, flammable, highly reactive, and explosive substances. These regulations apply to processes involving listed chemicals in amounts at or above threshold quantities. Employers subject to the PSM regulations are required to take specified steps, which include evaluating the hazards associated with the process, as well as developing and implementing operating procedures, employee training, emergency action plans, and compliance audits at least every 3 years, among other requirements.[59] Despite the hazards of ammonium nitrate, this chemical is not listed as one of the chemicals subject to these regulations. OSHA officials told us they did not know why ammonium nitrate was not included when the regulation was first issued.[60] According to the August 2013 chemical advisory, although ammonium nitrate is not covered by the PSM regulations, the production or use of ammonium nitrate may involve PSM- listed chemicals, and the manufacture of explosives, which may involve ammonium nitrate, is covered by the regulations. In the late 1990s, OSHA staff drafted a proposal for expanding PSM regulations to cover ammonium nitrate and other reactive chemicals, but it was not reviewed by agency policy officials and was never published in the Federal Register for public comment.[61]

In addition, retail facilities, which may include facilities that store and blend fertilizer for direct sale to end users, are exempt from OSHA's PSM regulations. In the preamble to the final rule for the PSM regulations, OSHA stated that retailers are not likely to store large quantities of hazardous chemicals, and that a large chemical release would be unlikely. While the facility in West, Texas stored large quantities of anhydrous ammonia, a chemical covered by the PSM regulations, OSHA officials told us that the PSM regulations would not apply to the facility because it was a retail outlet.

In addition, other chemical safety regulations issued by EPA do not apply to facilities with ammonium nitrate.[62] EPA's RMP regulations, issued in 1996 in response to a provision of the Clean Air Act Amendments of 1990, require covered chemical facilities to develop and implement a risk management program, but ammonium nitrate is not included on the list of chemicals that would trigger the requirements.[63] EPA's RMP regulations require facilities that handle more than threshold amounts of certain chemicals to implement a risk management program to guard against the release of chemicals into the air and surrounding environment. Covered facilities must develop their own risk management plans, and some facilities must also develop an emergency response program and conduct compliance audits, among other requirements. Covered facilities must also submit their risk management plans to EPA, including data on the regulated substances handled, and prepare a plan for a worst-case chemical release scenario.

Although EPA initially included high explosives in its list of regulated substances, which would include explosives grade ammonium nitrate, these explosives were subsequently removed from the list as a result of a legal settlement.[64] EPA officials also told us that fertilizer grade ammonium nitrate was not considered for its list for RMP because the agency had determined that it did not meet the criteria EPA established to implement the statute. Specifically, EPA officials told us that ammonium nitrate could have been included in the RMP regulations, but ammonium nitrate was not included because it was not considered a toxic or flammable chemical, which were among the criteria EPA used when the agency first developed the regulations.[65] Accordingly, ammonium nitrate is not a covered chemical and EPA inspectors do not review facilities' risk management plans for this chemical during their RMP inspections. In 2006, EPA conducted an on-site inspection of the West, Texas facility, but the inspection focused on anhydrous ammonia, not ammonium nitrate.

Under the Executive Order, OSHA and EPA Are Seeking Information on Expanding Regulation and Oversight of Ammonium Nitrate, but Have Not yet Proposed Any Regulatory Changes

In response to the August 2013 Executive Order on Improving Chemical Facility Safety and Security, OSHA and EPA, as part of the federal working group, have invited public comment on a wide range of policy options for overseeing the housing and handling of hazardous chemicals in the United States. Because they are still evaluating these options, the agencies have not issued any notices of proposed rulemaking. As directed by the Executive Order, in December 2013, OSHA issued a Request for Information on potential revisions to its PSM and related regulations, including its ammonium nitrate storage regulations.[66] OSHA's Request for Information also seeks public input on changing the agency's enforcement policy concerning the retailer exemption in the PSM regulations. In the Request for Information, OSHA states that "The West Fertilizer facility is not currently covered by PSM, however it is a stark example of how potential modernization of the PSM standard may include such facilities and prevent future catastrophe." In addition, as chair of one of the workgroups established to implement the Executive Order, OSHA solicited public input in January 2014 on federal policy options for improved chemical safety and security, including whether to expand OSHA's PSM regulations and EPA's RMP regulations to cover ammonium nitrate, among other options.[67] This solicitation also sought public input on whether federal agencies should examine the use of third party audits to promote safe storage and handling of ammonium nitrate. The solicitation defined third party audits as inspections conducted by independent auditors, retained by a chemical facility, who make process safety and regulatory compliance recommendations.[68] In an ongoing pilot project in selected states implemented in response to the Executive Order, federal agencies report improved coordination of inspections, such as sharing inspection schedules, cross-training inspectors, and inter-agency referrals of possible regulatory non-compliance.

SOME COUNTRIES REGULATE AND OVERSEE AMMONIUM NITRATE BY IMPOSING REQUIREMENTS ON FACILITIES, CONDUCTING INSPECTIONS, AND SUPPORTING INDUSTRY INITIATIVES TO PROMOTE COMPLIANCE

Other Countries' Approaches Include Risk Assessments and Restrictions on Where and How Ammonium Nitrate Can Be Stored

According to foreign officials and government documents, Canada and the three EU countries we contacted—France, Germany, and the United Kingdom—require facilities with specified quantities of ammonium nitrate, including fertilizer grade ammonium nitrate, to assess its risk and develop plans or policies to control the risks and mitigate the consequences of accidents.[69] Like the United States, these countries are members of the OECD, which has published best practices for managing the risks of chemical accidents.[70] The OECD publication includes guidance on preventing and mitigating the consequences of chemical accidents, preparedness planning, and land use planning, among other things.[71] For example, OECD's guidance recommends that regulatory authorities ensure that facilities with hazardous substances assess the range of possible accidents and require hazardous facilities to submit reports describing the hazards and the steps taken to prevent accidents.

With respect to assessing the risks of ammonium nitrate, according to Canadian officials and Canadian government documents, ammonium nitrate is regulated under the country's Environmental Emergency Regulations, which include risk management provisions. According to guidance published by Environment Canada, a federal-level regulatory agency, facilities that store 22 tons or more of ammonium nitrate must develop and implement an environmental emergency plan.[72] In developing an emergency plan, facilities are directed to analyze the risks posed during the storage and handling processes for certain chemicals and adopt practices to reduce the risks, taking into consideration the impact a chemical accident would have on the surrounding community.

According to information provided by EU officials, facilities in the 28 member countries of the EU with specific quantities of ammonium nitrate fertilizer are subject to the Seveso Directive, the EU legislation for facilities that use or store large quantities of certain toxic, explosive, and flammable substances, among other types of chemicals. [73] At a minimum, EU officials told us that EU member countries must comply with the Seveso Directive, although they have the option to adopt more stringent requirements. The legislation was adopted after a chemical accident in Seveso, Italy in 1976 that exposed thousands of people to the toxic chemical known as dioxin. Under the Seveso Directive, last updated in 2012, member countries are to require facilities with large amounts of ammonium nitrate fertilizer to notify the appropriate authority in their respective country, adopt a major accident prevention policy, and in some cases, develop a detailed safety report (see table 5).[74]

Some countries, such as France and the United Kingdom, have other requirements for notifying authorities about the types and quantities of chemicals at facilities, including certain types of ammonium nitrate. In the United Kingdom, officials told us that facilities with 28 tons or more of certain types of ammonium nitrate must notify the Health and Safety Executive or local authority and the fire authorities.[75] French officials said that facilities with more than 276 tons of ammonium nitrate fertilizer must notify local authorities about their holdings.[76]

Table 5. Selected Key Requirements and Corresponding Threshold Quantities in the European Union's Seveso III Directive for Facilities with Fertilizer Grade Ammonium Nitrate

Summary of directive requirement	Threshold quantity (in tons)
Notification Facilities are required to notify the appropriate authority in their country by submitting the names and quantities of chemicals present, activities performed, and details about neighboring establishments, including areas likely to increase the risk or consequences of a major accident.	1,378
Major accident prevention policy Facilities are required to document how they plan to prevent accidents and protect human health and the environment, including identifying and evaluating major hazards and planning for emergencies, among other activities, and submit the document to the appropriate authority in their country.	1,378
Safety report Facilities are required to produce a safety report demonstrating that major accident hazards and scenarios have been indentified and that measures have been taken to prevent such accidents, and send the report to the appropriate authority in their country.	5,512

Source: The Seveso III Directive and information provided by EU officials.

Note: The Seveso III Directive was adopted on July 4, 2012 and entered into force on August 13, 2012. EU member countries have until June 1, 2015 to implement the Seveso III Directive.

The selected countries we reviewed generally reported having more centralized land use policies that specify where facilities with large quantities of ammonium nitrate should be located. For example, EU officials explained that the Seveso Directive requires member countries to develop and implement land use policies. Through controls on the siting of new Seveso facilities and new developments in the vicinity of such facilities, such as transportation routes and residential areas, they told us, member countries' policies aim to limit the consequences of chemical accidents for human health and the environment. In the United Kingdom, officials told us that facilities intending to store more than 1,102 tons of ammonium nitrate must first receive permission from their local planning authority to do so for relevant ammonium nitrate materials. They explained that these local planning authorities consider the hazards and risks to people in surrounding areas and consult with the Health and Safety Executive prior to granting permission to such facilities.

Three of the countries we reviewed—France, Germany, and the United Kingdom—restrict the use of wood for storage purposes in certain instances, according to information and documents provided by relevant officials. EU officials told us that the Seveso Directive does not prescribe how chemicals, including ammonium nitrate, should be stored. EU countries have developed their own technical standards or rely on industry standards for storing and handling ammonium nitrate. For example, according to information provided by French officials, after several accidents involving ammonium nitrate fertilizer, the government in France launched a working group to update existing ammonium nitrate regulations, including storage and handling requirements. They described the most recent regulations in France, issued in 2010, which include updated fire resistance provisions for new and existing facilities banning or restricting the use of materials such as wood and asphalt flooring for storing ammonium nitrate. Specifically, according to documents provided by French officials, the regulations direct facilities not to store ammonium nitrate fertilizer in structures with wood walls or sides.[77] According to an official in Germany, strict storage requirements for using certain types of ammonium nitrate fertilizer have led many farmers to voluntarily use an alternative type of fertilizer, known as calcium ammonium nitrate.[78,79] For example, she explained that, in Germany, certain kinds of ammonium nitrate must be divided into quantities of 28 tons prior to storage, and quantities are separated by concrete walls. In addition, certain ammonium nitrate and ammonium nitrate-based preparations must be separated from combustible materials, for example by brick or concrete walls. Guidance in the United Kingdom also recommends that buildings for storing ammonium nitrate should be constructed of material that does not burn, such as concrete, bricks, or steel, as does the recent advisory in the United States published by OSHA, EPA, and ATF.

Foreign Oversight Approaches and Industry Initiatives Include Guidance on Safe Practices, Requirements for Routine Inspections, and Voluntary Third Party Audit Programs

Guidance on Safe Practices. In the countries we reviewed, government entities developed materials to help facilities with ammonium nitrate fertilizer comply with safety regulations.[80] For example, in the United Kingdom, the government published guidance on storing and handling ammonium nitrate that illustrates proper storage practices and is written in plain language. The United Kingdom also developed a checklist that facilities can use as a compliance tool to determine whether they are meeting safe storage requirements. In Canada, Environment Canada issued a guidance document in 2011 so that facilities covered by its Environmental Emergency Regulations, including facilities with certain types and amounts of ammonium nitrate, can better understand and comply with regulatory requirements.

The EU compiles information about chemical accidents and disseminates publications that include guidance on how facilities can prevent future incidents. Specifically, the EU has a system for reporting major accidents, including accidents involving ammonium nitrate, and tracks the information in a central database.[81] For example, as of January 2014, this database contained information on several incidents involving ammonium nitrate dating back to 1986. EU researchers use this information to develop semi-annual publications in order to facilitate the exchange of lessons learned from accidents for both industry and government regulators. Each publication focuses on a particular theme such as a specific substance, industry, or practice, and summarizes the causes of related accidents and lessons learned to help prevent future accidents. EU officials told us that the next publication will be issued in the summer of 2014 and will focus on the hazards of ammonium nitrate in part as a result of the explosion that occurred in West, Texas.

Routine Inspections. In the EU, member countries are required to inspect facilities with large quantities of chemicals covered by the Seveso Directive, which includes facilities with ammonium nitrate.[82] According to EU officials and documents, the EU's Seveso Directive requires covered facilities to be inspected either annually or once every 3 years, depending on the amount of hazardous chemicals a facility has—the greater the amount, the more frequent the inspection. EU officials also explained that member countries are required to report information to the European Commission every 3 years on how they are implementing the Seveso Directive requirements, including the number of facilities that have been inspected in their country.[83] According to a report published by the European Commission in June 2013, member countries reported in December 2011 that they had 10,314 covered facilities. According to the report, of those facilities to be inspected annually, 66 percent were inspected, on average, in 2011, and of those facilities to be inspected once every 3 years, 43 percent were inspected, on average, in 2011.[84]

Voluntary Initiatives and Third Party Audits. In the countries we reviewed, the fertilizer industry has actively promoted voluntary compliance with national safety requirements among facilities with ammonium nitrate fertilizer. For example, Fertilizers Europe, which represents the major fertilizer manufacturers in Europe, published guidance in 2007 for the storage and handling of ammonium nitrate-based fertilizers. This guidance recommends that buildings used to store ammonium nitrate-based fertilizers be constructed of non-readily combustible materials such as brick, concrete, or steel and that wood or other combustible materials be avoided, among other things.[85] Fertilizers Europe has also developed a compliance program that is a key requirement for membership, which consists of independent third party audits. As part of the program, it developed a self assessment tool for fertilizer manufacturers to use to identify gaps and possible improvements.

In the United Kingdom, the government and the fertilizer industry worked together in 2006 to develop a voluntary compliance program for facilities that manufacture and store fertilizers, among other activities, including ammonium nitrate-based fertilizers.[86]

According to a United Kingdom official, the government provided some of the initial funding for this initiative, and the voluntary compliance program is now self financed. Although the program was initially focused on fertilizer security, it has evolved over the years to also address fertilizer safety in the United Kingdom. As part of the voluntary compliance program, participating facilities carry out risk assessments. These facilities are audited annually by an independent audit team comprised of specialists to determine whether they comply with industry and government standards, including standards for safely storing and handling ammonium nitrate fertilizer.

Officials we interviewed in the United Kingdom told us that the government encourages and supports this industry initiative and that about 90 percent of facilities with ammonium nitrate in the United Kingdom, including those that have small quantities, are members of the voluntary program.[87] A United Kingdom official said, in his opinion, one would expect facilities participating in this industry initiative to be more likely to be found in compliance by the government when it conducts its own inspections. Furthermore, government officials, industry representatives, and program administrators meet twice a year to discuss how the program is being implemented and monitored.

CONCLUSION

Large quantities of ammonium nitrate are present in the United States, although the precise number of facilities with ammonium nitrate is not known. While incidents involving ammonium nitrate are rare, this chemical can react in ways that harm significant numbers of people and devastate communities. Facilities may be required, in certain circumstances, to report their chemical holdings to federal, state, and local authorities for security and emergency planning purposes. However, given the various reporting requirements and numerous reporting exemptions, some facilities may be uncertain about what to report to whom. Through the new Executive Order, federal agencies including DHS, EPA, and OSHA have the opportunity to work together on data sharing initiatives to help identify facilities with ammonium nitrate fertilizer. Such data sharing could help federal agencies identify facilities that are not complying with their regulations and enable OSHA to target high risk facilities with ammonium nitrate for inspection. Without improved coordination among the various federal and state agencies that collect data on facilities that store potentially hazardous chemicals, identifying facilities with ammonium nitrate for purposes of increasing awareness of the hazards and improving regulatory compliance will remain a challenge.

Although OSHA has requirements for storing ammonium nitrate fertilizer in its Explosives and Blasting Agents regulations that could reduce the likelihood of an explosion, OSHA has done little to ensure that the fertilizer industry, which is one of the primary users of ammonium nitrate, understands how to comply with its existing regulations. The August 2013 chemical advisory and OSHA's February 2014 letter to facilities help clarify how OSHA's Explosives and Blasting Agents regulations apply to fertilizer facilities. However, without additional action by OSHA to promote awareness of how to comply with its regulations, fertilizer facilities may not know whether their practices are in compliance with OSHA's existing ammonium nitrate storage regulations or if changes need to be made. Moreover, unless OSHA takes steps to leverage additional resources to support its enforcement efforts, whether through enhanced targeting or coordination with other agencies or outside parties, beginning with encouraging voluntary compliance with ammonium nitrate regulations through various industry initiatives, it will not know the extent to which dangerous conditions at some facilities may continue to exist.

While much can be achieved under current regulations, OSHA and EPA's regulations contain gaps with respect to ammonium nitrate that may allow unsafe facilities to operate and poor planning to persist. OSHA has not significantly changed its ammonium nitrate storage regulations since they were issued in 1971, which means that fertilizer facilities may be adhering to outdated practices. For example, other countries we reviewed have revisited and updated their ammonium nitrate regulations and the National Fire Protection Association is considering making changes to its ammonium nitrate storage standards as a result of the explosion in West, Texas. In addition, as a result of incidents involving ammonium nitrate abroad, countries in the European Union and Canada require facilities to assess the risks of working with ammonium nitrate fertilizer, and the European Union requires member countries to routinely inspect facilities that have very large quantities of it. These approaches offer examples of how the risks of ammonium nitrate can be managed. Although increased regulation may be more burdensome to industry, without some means of ensuring that high risk facilities plan for and manage the risks associated with ammonium nitrate, such facilities may not be prompted to adequately address the risks the chemical creates for workers and neighboring communities.

RECOMMENDATIONS FOR EXECUTIVE ACTION

1. To improve federal oversight of facilities with ammonium nitrate, we recommend that the Secretary of Labor, the Administrator of EPA, and the Secretary of Homeland Security, as part of their efforts as members of the Chemical Facility Safety and Security Working Group established by the Executive Order issued in August 2013, develop and implement methods of improving data sharing among federal agencies and with states.
2. We also recommend that the Secretary of Labor direct the Assistant Secretary for Occupational Safety and Health to take the following three actions:
   - Extend OSHA's outreach to the fertilizer industry. For example, OSHA could work with the fertilizer industry to develop and disseminate informational materials related to storage of ammonium nitrate.
   - Take steps to identify high risk facilities working with ammonium nitrate and develop options to target them for inspection.

- Consider updating regulations for the storage of ammonium nitrate taking into consideration, as appropriate, other related standards and current practices.

3. To strengthen federal oversight of facilities with ammonium nitrate, we recommend that the Secretary of Labor and the Administrator of EPA direct OSHA and EPA, respectively, to consider revising their related regulations to cover ammonium nitrate and jointly develop a plan to require high risk facilities with ammonium nitrate to assess the risks and implement safeguards to prevent accidents involving this chemical.

AGENCY COMMENTS AND OUR EVALUATION

We provided a draft of this report to the Administrator of EPA, the Secretary of Homeland Security, and the Secretary of Labor for review and comment. We received written comments from EPA, DHS, and OSHA, EPA, DHS, and OSHA agreed with our recommendation that the agencies improve data sharing and described their current efforts to address this issue as part of their implementation of the Executive Order on Improving Chemical Facility Safety and Security. The agencies stated that a status report by the Executive Order Working Group, which will be submitted to the President by the end of May, 2014, will include proposals for enhancing data sharing among federal agencies and with states.

OSHA agreed with our recommendation that the agency conduct additional outreach to the fertilizer industry, stating that additional outreach efforts will be identified in the Executive Order status report and that these efforts should help the fertilizer industry understand OSHA's safety requirements and industry best practices. OSHA also agreed with our recommendation that the agency target high risk facilities for inspection, stating that the agency is evaluating options for targeting high risk fertilizer facilities for inspection.

OSHA and EPA agreed with our recommendation that the agencies consider revising their regulations to cover ammonium nitrate. OSHA is currently reviewing public comments submitted in response to a Request for Information on a proposed revision to the agency's Process Safety Management and Prevention of Major Chemical Accidents regulations and the a request for public input on issues associated with Section 6 of the Executive Order, which addresses Policy, Regulation, and Standards Modernization.

EPA stated that the agency will be publishing a Request for Information seeking public input on its proposed revision to process safety and risk management issues relevant to its Risk Management Program regulations, including coverage of ammonium nitrate. In addition, EPA, DHS, and OSHA provided technical comments, which we have incorporated as appropriate. We also provided portions of the draft report related to each of the four countries we reviewed to relevant officials from each country, and incorporated their technical comments, as appropriate.

Revae Moran, Director
Education, Workforce, and Income Security Issues

End Notes

[1] *Hearing on Oversight of Federal Risk Management and Emergency Programs to Prevent and Address Chemical Threats, Including the Events Leading up to the Explosions in West, Texas and Geismar, Louisiana, Before the Senate Comm. on Environment and Public Works*, 113th Cong. 1st Sess., June 27, 2013 (statement of Rafael Moure-Eraso, Chairman, Chemical Safety and Hazard Investigation Board (Chemical Safety Board). The Chemical Safety Board is an independent federal safety board charged with investigating chemical accidents.

[2] Association of American Plant Food Control Officials and The Fertilizer Institute, *Commercial Fertilizers 2012*, Columbia, Missouri:. This publication reports fertilizer consumption data submitted by state fertilizer control offices. The consumption data include total sales or shipments of fertilizer for farm and non-farm use by state.

[3] U.S. Census Bureau, *Current Industrial Reports, Fertilizers and Related Chemicals – 2010*, MQ325B(10)-5, June 2011.

[4] Improving Chemical Facility Safety and Security, Exec. Order No. 13,650, 78 Fed. Reg. 48,029 (Aug. 7, 2013).

[5] In this report, we use the term facility to mean any fixed site where hazardous chemicals are present, which can include chemical manufacturers, distributors, and farm supply retailers. The term facility may be defined differently for regulatory purposes.

[6] DHS requires facilities to report if they possess certain chemicals at or above its screening threshold quantities. This may include facilities that manufacture, process, use, store, or distribute these chemicals.

[7] These four states accounted for about 55 percent of U.S. ammonium nitrate fertilizer consumption in 2012. Alabama represents about 10 percent, Missouri represents about 19 percent, Tennessee represents about 18 percent, and Texas represents about 8 percent. Source: Commercial Fertilizers 2012 report published by the Association of American Plant Food Control Officials and The Fertilizer Institute.

[8] These data are collected pursuant to the Emergency Planning and Community Right-to- Know Act of 1986 (EPCRA). State responses to our data requests appeared, in part, to reflect differing interpretations of EPCRA. Officials from Missouri and Tennessee said that they would provide data only in response to a written request about specific facilities, citing an EPCRA provision on availability of data to the public. Because a complete list of facilities with ammonium nitrate was not readily available, we were unable to ask for data by facility.

[9] DHS's Customs and Border Protection agency collects real time data on shipments of products to and from the United States as part of its efforts to facilitate international trade and protect national security.

[10] In this review, we sought to identify federal regulations that apply to ammonium nitrate used as a fertilizer. Although the regulations we identified may also apply to ammonium nitrate used for other purposes (for example, as a blasting agent), additional federal regulations may apply in these contexts that are not discussed in this report.

[11] According to the Chemical Safety Board, a granular solid form of ammonium nitrate was stored at the West, Texas facility. *Senate Hearing on Chemical Threats* (June 27, 2013) (statement of CSB Chairman Rafael Moure-Eraso). Fertilizer sales data published by the Department of Agriculture suggest that solid ammonium nitrate fertilizer represents about 3 percent of all types of fertilizer sold in the United States and that ammonium nitrate fertilizer sales have generally declined in recent years.

[12] The Fertilizer Institute is a national organization representing producers, importers, retailers, and others involved in the fertilizer industry.

[13] Products containing ammonium nitrate can vary in their composition and chemical properties, depending on the purpose for which they will be used, such as a fertilizer or as an explosive. Different types of ammonium nitrate may be subject to different regulatory requirements, as discussed later in this report.

[14] Information about the hazards of ammonium nitrate can be found in the International Chemical Safety Card for Ammonium Nitrate published by the National Institute for Occupational Safety and Health, which is part of the Department of Health and Human Services' Centers for Disease Control and Prevention.

[15] See GAO, *Combating Terrorism: State Should Enhance Its Performance Measures for Assessing Efforts in Pakistan to Counter Improvised Explosive Devices,* GAO-12-614 (Washington, D.C.: May 15, 2012).

[16] *Senate Hearing on Chemical Threats* (June 27, 2013) (statement of CSB Chairman Rafael Moure-Eraso).

[17] Pub. L. No. 91-596, 84 Stat. 1590 (codified as amended at 29 U.S.C. §§ 553, 651-78). OSHA's regulations on hazardous materials may be found in subpart H of 29 C.F.R. pt. 1910. OSHA's regulations apply to private sector workplaces and some federal government workplaces. In this report, we focused on regulations that apply to private sector workplaces.

[18] Pub. L. No. 101-549, §§ 301, 304, 104 Stat. 2399, 2563-74, 2576-77.

[19] See 29 C.F.R. § 1910.119 and app. A (OSHA's regulation on process safety management of highly hazardous chemicals) and 40 C.F.R. pt. 68 (EPA's risk management program regulations).

[20] Pub. L. No. 99-499, tit. III, 100 Stat. 1613, 1728-58 (codified at 42 U.S.C. §§ 11001-50).

[21] 42 U.S.C. § 11022. EPA's regulations implementing sections 311 and 312 of EPCRA, pertaining to hazardous chemical reporting, are found at 40 C.F.R. pt. 370. As discussed later in this report, according to an August 2013 chemical advisory issued by the Bureau of Alcohol, Tobacco, Firearms and Explosives (ATF), EPA, and OSHA, ammonium nitrate is considered a hazardous chemical subject to the EPCRA reporting provisions. However, EPCRA exempts any substance "to the extent it is used in routine agricultural operations or is a fertilizer held for sale by a retailer to the ultimate customer." 42 U.S.C. § 11021(e)(5). According to the advisory, this exemption applies only to ammonium nitrate retailers, not to manufacturers or wholesalers; any ammonium nitrate that is mixed or formulated with other chemicals by facilities is not covered by the exemption.

[22] 6 C.F.R pt. 27 and app. A. DHS established the CFATS program in response to a requirement in its annual appropriations. See Department of Homeland Security Appropriations Act, 2007, Pub. L. No. 109-295, § 550, 120 Stat. 1355, 1388-89 (2006). DHS issued the CFATS regulations in 2007 and the list initially included 322 chemicals of interest and the screening threshold quantities for each chemical. Additional requirements may apply to specific

facilities based on the information provided or other factors. For more information on the CFATS program see: GAO, *Critical Infrastructure Protection: DHS Needs to Improve Its Risk Assessments and Outreach for Chemical Facilities*, GAO-13-801T (Washington, D.C.: Aug. 1, 2013); GAO, *Critical Infrastructure Protection: DHS Efforts to Assess Chemical Security Risk and Gather Feedback on Facility Outreach Can Be Strengthened*, GAO-13-353 (Washington, D.C.: Apr. 5, 2013); and GAO, *Critical Infrastructure Protection: Observations on DHS Efforts to Identify, Prioritize, Assess, and Inspect Chemical Facilities*, GAO-14-365T (Washington, D.C.: Feb. 27, 2014).

[23] North American Industry Classification System (NAICS) industry codes are used to classify the industry that best describes the facilities that report to DHS.

[24] Pursuant to its authority under 6 C.F.R. § 27.210(c), DHS has extended the deadline for submitting CFATS reports until further notice for certain agricultural production facilities, such as farms, ranches, turfgrass growers, golf courses, nurseries, and public and private parks. See Notice to Agricultural Facilities About Requirement To Complete DHS' Chemical Security Assessment Tool, 73 Fed. Reg. 1640 (Jan. 9, 2008).

[25] DHS's CFATS regulations provide that in calculating whether a facility possesses a threshold amount of a chemical that poses a theft or diversion risk, the facility shall only include those chemicals that are in transportation packaging as defined by Department of Transportation regulations. 6 C.F.R. § 27.203(c). DHS considers ammonium nitrate fertilizer a chemical of interest because it can be stolen or otherwise diverted to make explosives.

[26] Department of Homeland Security Appropriations Act, 2007, Pub. L. No. 109-295, § 550(a), 120 Stat. 1355, 1388-89 (2006), 6 C.F.R. § 27.110(b).

[27] ATF collects data on individuals that apply for federal explosives licenses and permits, which may include individuals working with ammonium nitrate.

[28] The OSH Act allows states to take responsibility for operating their own occupational safety and health programs under state plans approved by OSHA. To receive approval, state plans must meet certain criteria specified in the OSH Act, including the development and enforcement of state standards that are at least as effective as the federal standards. See generally 29 U.S.C. § 667, 29 C.F.R. pts. 1902, 1952, and 1956. Under the OSH Act, "state" is defined to include the District of Columbia, Puerto Rico, the Virgin Islands, American Samoa, Guam, and the Trust Territory of the Pacific Islands. See 29 U.S.C. § 652(7).

[29] OSHA does not enforce standards for state and local public-sector workplaces because the OSH Act does not apply to state and local government employers. 29 U.S.C. § 652(5). States that choose to operate their own state-run programs are required to cover state and local government workers. 29 U.S.C. § 667(c)(6). Five states have state plans that only include state and local government workers; OSHA provides enforcement for the private sector in those states.

[30] Under the Clean Air Act, EPA is authorized to delegate its implementation and enforcement authority of section 112 (including the RMP program) to states, provided the state standards are no less stringent than EPA's. 42 U.S.C. § 7412(l), 40 C.F.R. §§ 63.90-63.99. According to EPA officials, the nine states to which EPA has delegated this authority are: Delaware, Florida, Georgia, Mississippi, New Jersey, North Carolina, North Dakota, Ohio, and South Carolina. The five counties to which EPA has delegated this authority are: Buncombe, North Carolina; Forsyth, North Carolina; Mecklenburg, North Carolina; Jefferson, Kentucky; and Allegheny, Pennsylvania.

[31] In addition to the specific provisions focused on ammonium nitrate, the Executive Order also addresses other hazardous chemicals more generally.

[32] The 34 OECD member countries are Australia, Austria, Belgium, Canada, Chile, Czech Republic, Denmark, Estonia, Finland, France, Germany, Greece, Hungary, Iceland, Ireland, Israel , Italy, Japan, Korea, Luxembourg, Mexico, Netherlands, New Zealand, Norway, Poland, Portugal, Slovak Republic, Slovenia, Spain, Sweden, Switzerland, Turkey, United Kingdom, and the United States. See *OECD Guiding Principles for Chemical Accident*

Prevention, Preparedness and Response: Guidance for Industry (including Management and Labour), Public Authorities, Communities, and other Stakeholders (OECD 2003).

[33] Under section 312 of EPCRA and EPA's regulations, facilities with 10,000 pounds or more of ammonium nitrate generally must submit an annual chemical inventory report to their designated state and local authorities. 42 U.S.C. § 11022, 40 C.F.R. § 370.10(a)(2)(i).The designated authorities are the state emergency response commission, the local emergency planning committee, and the local fire department. A facility is required to submit these reports if (1) it is required to prepare a material safety data sheet (now called a safety data sheet) for a hazardous chemical as defined by OSHA's Hazard Communication regulations, 29 C.F.R. § 1910.1200, and (2) the amount of the hazardous chemical meets or exceeds the threshold set by EPA's regulations. For most hazardous chemicals that are not on EPA's list of Extremely Hazardous Substances, the reporting threshold is 10,000 pounds or more. According to the chemical advisory issued by ATF, EPA, and OSHA in August 2013, ammonium nitrate is not considered an Extremely Hazardous Substance, but it is considered a hazardous chemical under OSHA's Hazard Communication regulations and is therefore subject to the EPCRA provisions. However, EPCRA exempts any substance "to the extent it is used in routine agricultural operations or is a fertilizer held for sale by a retailer to the ultimate customer." 42 U.S.C. § 11021(e)(5). According to the advisory, this exemption applies only to ammonium nitrate retailers, not to manufacturers or wholesalers; any ammonium nitrate that is mixed or formulated with other chemicals by facilities is not covered by the exemption.

[34] We compared the facility names and zip codes of facilities reporting to the state to the facilities that reported to DHS. Because of differences in reporting requirements, the differences in the number of facilities reporting to DHS and the number reporting to the states does not necessarily indicate noncompliance with the requirements.

[35] In Texas, facility owners must register with the Office of the Texas State Chemist to produce, store, or sell ammonium nitrate, and there is no minimum threshold amount of ammonium nitrate that applies to this state requirement. See Tex. Agric. Code Ann. §§ 63.151-63.157. This requirement applies to ammonium nitrate that contains more than 33 percent nitrogen, as well as solid fertilizer containing ammonium nitrate, if the fertilizer's nitrogen content from the ammonium nitrate is at least 28 percent of the fertilizer by weight. Facilities are required to keep records of the sale of ammonium nitrate and provide the records upon request to the State Chemist and other state agencies.

[36] We did not find any other state agencies in Alabama that required reporting of ammonium nitrate holdings similar to the Office of the Texas State Chemist.

[37] We compared the facility names and the city names used by companies that import and export ammonium nitrate to the facilities that reported to DHS.

[38] We counted any facility that imported or exported products with "ammonium nitrate" listed as part of the product description.

[39] We identified imports or exports of "ammonium nitrate," but the federal trade data did not provide the actual chemical composition of the fertilizer; therefore, we could not determine whether these facilities were potentially subject to CFATS reporting requirements.

[40] See 42 U.S.C. § 11021(e)(5) and Notice to Agricultural Facilities About Requirement To Complete Chemical Security Assessment Tool Top-Screen, 73 Fed. Reg. 1640 (Jan. 9, 2008).

[41] DHS's CFATS regulations provide that, in calculating whether a facility possesses a threshold amount of a chemical that poses a theft or diversion risk, the facility shall only include those chemicals that are in transportation packaging as defined by DOT regulations. 6 C.F.R. § 27.203(c). DHS considers ammonium nitrate fertilizer a chemical of interest because it can be stolen or otherwise diverted to make explosives.

[42] See 42 U.S.C. § 11022(e)(3).

[43] In this review, we sought to identify federal regulations that apply to ammonium nitrate used as a fertilizer. Although the regulations we identified may also apply to ammonium nitrate

used for other purposes (for example, as a blasting agent), additional federal regulations may apply in these contexts that are not discussed in this report.

[44] 29 C.F.R. § 1910.109(i). These provisions apply to the storage of ammonium nitrate in the form of crystals, flakes, grains, or prills including fertilizer grade, dynamite grade, nitrous oxide grade, technical grade, and other mixtures containing 60 percent or more ammonium nitrate by weight, but do not apply to blasting agents.

[45] NFPA is an independent nonprofit organization that convenes technical committees to develop national codes and standards intended to minimize the possibility and effects of fire and other risks. NFPA codes and standards are developed by consensus by committees composed of representatives from the government, industry, fire associations, and other organizations. Unlike OSHA's regulations, consensus standards are voluntary.

[46] All flooring in storage and handling areas must be of noncombustible material or protected against impregnation by ammonium nitrate. 29 C.F.R. § 1910.109(i)(2)(iii)(d). For bulk storage of ammonium nitrate, wooden bins protected against impregnation by ammonium nitrate are permissible. 29 C.F.R. § 1910.109(i)(4)(ii)(b).

[47] 29 C.F.R. § 1910.109(i)(2)(iii)(e).

[48] Environmental Protection Agency (EPA), Occupational Safety and Health Administration (OSHA), and the Bureau of Alcohol, Tobacco, Firearms and Explosives (ATF) *Chemical Advisory: Safe Storage, Handling, and Management of Ammonium Nitrate,* EPA 555-S-13-001 (Washington, D.C.: August 2013).

[49] 29 C.F.R. § 1910.1200. The regulations do not include a list of chemicals and threshold amounts that would trigger application of the regulations. Rather, the regulations apply to any chemical which is known to be present in the workplace in such a manner that employees may be exposed under normal conditions of use or in a foreseeable emergency, subject to certain exceptions. In its 2012 revisions to the Hazard Communication regulations, OSHA changed the name of material safety data sheets to safety data sheets. See Hazard Communication, 77 Fed. Reg. 17,574 (Mar. 26, 2012).

[50] Section 21 of the OSH Act requires OSHA to establish programs to educate and train employers and employees in the recognition, avoidance, and prevention of unsafe or unhealthful working conditions, and to consult with and advise employers, employees, and organizations representing employers and employees as to effective means of preventing occupational injuries and illnesses. 29 U.S.C. § 670(c).

[51] Explosives, Part III, 72 Fed. Reg. 18,792 (Apr. 13, 2007).

[52] Manufacturers are required to develop safety data sheets for users of their hazardous chemical products, including ammonium nitrate fertilizer, under OSHA's Hazard Communication regulations. Although not required by OSHA, safety data sheets typically include a regulatory information section.

[53] 29 U.S.C. § 657.

[54] OSHA officials said the agency considers facilities that are classified as the highest risk category in EPA's RMP database to likely be subject to OSHA's PSM regulations. OSHA estimates that about 8,480 facilities are covered by its PSM regulations. As previously stated, although OSHA acquires data on facilities with hazardous chemicals from EPA's RMP database, that database does not include information on facilities with ammonium nitrate.

[55] In October 2013, OSHA cited the West, Texas facility and proposed penalties of $118,300 for violations of its ammonium nitrate storage regulations and other OSHA regulations. According to OSHA officials, these citations were contested by the employer and were pending before the Occupational Safety and Health Review Commission. OSHA last inspected the West, Texas facility in 1985. At that time, OSHA fined the facility $30 for violations of its regulations on storage and handling of anhydrous ammonia. Anhydrous ammonia is a colorless gas with a pungent, suffocating odor that can be compressed to make a liquid fertilizer. Anhydrous ammonia is considered a high health hazard because it is

corrosive to the skin, eyes, and lungs. Anhydrous ammonia is also flammable and can explode under certain conditions.

[56] See, for example, Departments of Labor, Health and Human Services, and Education, and Related Agencies Appropriations Act, 2012, Pub. L. No. 112-74, div. F, tit. I, 125 Stat. 786, 1059-60 (2011), which provides that, subject to certain exceptions, no appropriated funds shall be used to enforce any regulation under the OSH Act "with respect to any employer of 10 or fewer employees who is included within a category having a[n] ... occupational injury and illness rate ...less than the [most recent] national average," as published by the Department of Labor's Bureau of Labor Statistics. The exceptions include, among others, inspections for health hazards and unplanned inspections (such as those conducted in response to employee complaints or serious accidents). The Department of Labor's Bureau of Labor Statistics estimates workplace injury and illness rates by industry using North American Industry Classification System industry codes. To identify which industries are subject to OSHA's enforcement exemption, OSHA periodically updates one of its enforcement directives to list the most current North American Industry Classification System codes for each industry with an average workplace injury and illness rate below the national average.

[57] For ammonium nitrate meeting the definition of ammonium nitrate regulated by the Department of Transportation as a Division 1.1 Explosive, facilities are generally required to report to DHS if they have 400 pounds or more contained in transportation packaging (if not contained in transportation packaging, the threshold is 5,000 pounds or more). For ammonium nitrate commonly used as a fertilizer, facilities are generally required to report to DHS if they have 2,000 pounds or more contained in transportation packaging.

[58] 29 C.F.R. § 1910.119. OSHA's list of highly hazardous chemicals is found at 29 C.F.R. § 1910.119, app. A.

[59] A process means any activity involving a highly hazardous chemical including any use, storage, manufacturing, handling, or the on-site movement of such chemicals, or a combination of these activities.

[60] According to OSHA officials, ammonium nitrate met the criteria the agency used to develop the list of chemicals subject to the PSM regulations, but ammonium nitrate was not included in the final regulations and the agency could not find documentation that would explain why. The preamble to the PSM final rule states that the agency decided to include substances with the two highest or most dangerous reactivity ratings from NFPA's Hazardous Chemicals Data document 49 (substances rated 3 or 4 by NFPA). Process Safety Management of Highly Hazardous Chemicals; Explosives and Blasting Agents, 57 Fed. Reg. 6356, 6364 (Feb. 24, 1992). Ammonium nitrate has a reactivity rating of 3 from NFPA, but was not included in the list of chemicals subject to the PSM requirements in the final rule.

[61] A description of proposed plans to amend the PSM regulations to expand the chemicals covered appeared on OSHA's spring 1997 regulatory agenda and was removed as of the fall 2001 agenda. OSHA commissioned a study that was completed in 2000 by CONSAD Research Corporation, which included a preliminary chemical and industry profile and an economic analysis of the impacts of adding reactive chemicals to the scope of the PSM regulations, including ammonium nitrate.

[62] According to Chemical Safety Board officials, in 2002, the CSB recommended that OSHA and EPA expand the PSM and RMP regulations to include reactive chemicals. Ammonium nitrate is considered a reactive chemical. For more information, see U.S. Chemical Safety and Hazard Investigation Board, Hazard Investigation: Improving Reactive Hazard Management, Report No. 2001-01-H (Washington D.C.: October 2002).

[63] 40 C.F.R. pt. 68. EPA's list of regulated toxic and flammable substances is found at 40 C.F.R. § 68.130.

[64] As part of settlement agreements resolving legal challenges to EPA's list by members of the explosives industry, EPA agreed to propose de-listing high explosives and the industry

members agreed to develop and implement certain safety practices and to dismiss their legal challenge if the agency finalized the de-listing. EPA concluded that existing regulations and these industry practices were adequate to protect the public and finalized the de-listing. The industry members withdrew their legal challenge. List of Regulated Substances and Thresholds for Accidental Release Prevention; Amendments, 63 Fed. Reg. 640, 641 (Jan. 6, 1998).

[65] When it established its list of regulated substances, EPA included substances that met specified criteria for toxic, flammable, and explosive substances. For explosives, EPA selected substances that were given a certain explosive classification by the Department of Transportation. List of Regulated Substances and Thresholds for Accidental Release Prevention, 59 Fed. Reg. 4478 (Jan. 31, 1994). The Department of Transportation does not classify ammonium nitrate fertilizer as an explosive. 49 C.F.R. § 172.101.

[66] Process Safety Management and Prevention of Major Chemical Accidents, 78 Fed. Reg. 73,756 (Dec. 9, 2013).

[67] These policy options have been published to OSHA's website and public comments may be obtained through the website www.regulations.gov.

[68] In December 2012, the Administrative Conference of the United States, an independent federal agency dedicated to improving the regulatory process, published a recommendation on agency use of third-party programs to assess regulatory compliance. The recommendation refers to existing third-party inspection programs in which regulated entities generally contract with and pay third parties to carry out activities such as facility inspections. Regulatory agencies then adopt new roles in coordinating and overseeing these third parties. The Administrative Conference of the United States recommended that federal agencies consider various factors, such as resources and incentives to participate in a third-party inspection program, when deciding whether or not to develop such a program. It also acknowledges that certain statutory or other legal restrictions may preclude an agency from using third parties to conduct inspections and other compliance assistance activities. Adoption of Recommendations, 78 Fed. Reg. 2939, 2941-43 (Jan. 15, 2013).

[69] We did not conduct an independent legal analysis to verify the information provided about the laws, regulations, or policies of the foreign countries selected for this study.

[70] The OECD is an intergovernmental organization in which representatives meet to coordinate and harmonize policies, discuss issues of mutual interest, and respond to international concerns. Currently, there are 34 member countries.

[71] *OECD Guiding Principles for Chemical Accident Prevention, Preparedness and Response: Guidance for Industry (including Management and Labour), Public Authorities, Communities, and other Stakeholders* (OECD 2003).

[72] This includes mixtures that are 60 percent ammonium nitrate by weight and that are in solid form and mixtures that are 81 percent ammonium nitrate by weight and that are in liquid form.

[73] Currently, the Seveso Directive specifically covers four different types of ammonium nitrate, and reporting requirements for facilities vary depending on the quantity of ammonium nitrate they hold. The four types of ammonium nitrate covered are described in the Seveso Directive as: (1) ammonium nitrate fertilizers capable of self-sustaining decomposition, (2) fertilizer grade ammonium nitrate, (3) technical grade ammonium nitrate, and (4) "off-specs" material and fertilizers not fulfilling the detonation test. Threshold quantities vary depending on the type of ammonium nitrate. Fertilizer grade ammonium nitrate is defined in the Seveso Directive as straight ammonium nitrate-based fertilizers and ammonium nitrate-based compound/composite fertilizers that contain certain percentages of nitrogen from ammonium nitrate by weight. For more specific information, see Annex I of *Directive 2012/18/EU of the European Parliament and of the Council on the Control of Major-Accident Hazards Involving Dangerous Substances, Amending and Subsequently Repealing Council Directive 96/82/EC* (July 4, 2012). For purposes of this report, we focus on examples involving fertilizer grade ammonium nitrate.

[74] According to information provided by EU officials, the EU began regulating ammonium nitrate fertilizer in 1982. Subsequent to the adoption of the original Seveso Directive in 1982, there have been two replacement directives. Seveso II was adopted in 1996 and introduced requirements related to emergency planning and land use planning, among other revisions. Seveso II was amended in 2003 and changes were made to the descriptions of the ammonium nitrate categories and thresholds modifying the criteria for which facilities are covered under the Directive, among other changes. These changes were made based on an analysis of findings from the 2001 accident in Toulouse, France. Seveso III was adopted on July 4, 2012 and entered into force on August 13, 2012. EU member countries have until June 1, 2015 to implement the Seveso III Directive. Revisions include stricter standards for inspections to ensure more effective enforcement, and stricter requirements for providing information to the public, particularly those likely to be affected by a major accident, among other changes. This information was provided and/or reviewed by EU officials, for more details, see European Union, *Directive 2012/18/EU of the European Parliament and of the Council on the Control of Major- Accident Hazards Involving Dangerous Substances, Amending and Subsequently Repealing Council Directive 96/82/EC* (July 4, 2012).

[75] According to United Kingdom officials we interviewed, these requirements apply to grades of ammonium nitrate that are classified as oxidizers. The relevant regulations that require facilities to notify authorities are The Dangerous Substances (Notification and Marking of Sites) Regulations 1990. These regulations are primarily intended to alert fire authorities to any special firefighting hazards likely to exist at facilities. The Health and Safety Executive is a non-departmental United Kingdom government body.

[76] More specifically, according to officials, facilities with more than 276 tons but less than 551 tons of ammonium nitrate fertilizer that is more than 28 percent nitrogen from ammonium nitrate by weight and complies with EU standards, including passing a detonation resistance test, are required to notify local authorities. According to a French official, facilities with 11 tons or more of "off-spec" ammonium nitrate that does not comply with certain EU standards are classified as Seveso facilities.

[77] In this example, ammonium nitrate fertilizer refers to solid straight and compound fertilizers with specific percentages of nitrogen from ammonium nitrate by weight.

[78] The German official told us that German regulations apply to the storage, filling, and in- house transport of ammonium nitrate and ammonium nitrate-based preparations. In Germany, ammonium nitrate and ammonium nitrate-based preparations are classified into five groups based on their hazardous properties. For more information, see Germany's Hazardous Substances Ordinance, Federal Institute for Occupational Safety and Health, last amended July 15, 2013.

[79] According to an official we interviewed from an international fertilizer association, using calcium ammonium nitrate, which is a mixture of ammonium nitrate with limestone and/or dolomite, entails some incremental cost associated with the additional weight of the material added to the ammonium nitrate. According to a German official, under normal storage conditions, calcium ammonium nitrate fulfilling certain requirements is considered a safer fertilizer than straight ammonium nitrate fertilizer in terms of preventing accidental detonation, and large protection distances are used for straight ammonium nitrate fertilizer in case of accidental detonation. However, calcium ammonium nitrate fertilizer can still be used to make weapons, such as improvised explosive devices.

[80] OECD's *Guiding Principles for Chemical Accident Prevention, Preparedness and Response* directs public authorities to provide facilities with clear, easy to understand guidance on how regulatory requirements can be met.

[81] For more information, see the European Commission's Major Accident Reporting System https://emars.jrc.ec.europa.eu/.

[82] OECD's *Guiding Principles for Chemical Accident Prevention, Preparedness and Response* directs public authorities to ensure safety requirements are met through appropriate

inspection and enforcement measures, such as periodically inspecting safety performance in hazardous facilities.

[83] Under the new Seveso III Directive, member countries are required to report information to the European Commission every 4 years.

[84] These facilities are not just facilities with ammonium nitrate, but include facilities with more than threshold amounts of all of the chemicals covered by Seveso. For more information see European Commission, *Report on the Application in the Member States of Directive 96/82/EC on the control of major-accident hazards involving dangerous substances for the period 2009-2011* (Brussels, June 2, 2013).

[85] European Fertilizer Manufacturers' Association, *Guidance for the Storage, Handling and Transportation of Solid Mineral Fertilizers* (April 2007).

[86] The voluntary compliance program in the United Kingdom is known as the Fertilizer Industry Assurance Scheme.

[87] The voluntary program in the United Kingdom does not apply to end users, such as farms.

In: Ammonium Nitrate Facilities
Editor: Allegra Ward
ISBN: 978-1-63321-716-4

Chapter 2

REGULATION OF FERTILIZERS: AMMONIUM NITRATE AND ANHYDROUS AMMONIA*

Dana A. Shea, Linda-Jo Schierow and Scott D. Szymendera

SUMMARY

The explosion on April 17, 2013, at the West Fertilizer Company fertilizer distribution facility in West, TX, has led to questions about the oversight and regulation of agricultural fertilizer. Facilities holding chemicals must comply with regulations attempting to ensure occupational safety, environmental protection, and homeland security. In addition to federal regulation requiring reporting and planning for ammonium nitrate and anhydrous ammonia, most state and some local governments have laws and regulations regarding the handling of either or both of these chemicals.

The West Fertilizer Company possessed a variety of agricultural chemicals at its retail facility, but policy interest has focused on two chemicals: ammonium nitrate and anhydrous ammonia. Ammonium nitrate is a solid that can be used as a fertilizer, a use that generally occurs without

* This is an edited, reformatted and augmented version of a Congressional Research Service publication R43070, prepared for Members and Committees of Congress dated July 31, 2013.

incident. In combination with a fuel source and certain conditions, such as added heat or shock, confinement, or contamination, ammonium nitrate can pose an explosion hazard. Such accidents have rarely occurred, but have historically had high impacts. For example, the ammonium nitrate explosion in 1947 in Texas City, TX, where two ships carrying ammonium nitrate coated in wax and stored in paper bags caught fire and exploded, destroyed the entire dock area, including numerous oil tanks, dwellings, and business buildings. The bomb used in 1995 to attack the Murrah Federal Building in Oklahoma City, OK, contained ammonium nitrate as a component of its explosives.

Anhydrous ammonia has a variety of uses, including as an agricultural fertilizer. Many agricultural retailers store and use anhydrous ammonia. In contrast with ammonium nitrate, anhydrous ammonia is a gas more generally viewed as a threat from its inhalation toxicity. It is regulated to prevent release of the chemical into the atmosphere where it might travel as a cloud and impact workers and the surrounding environment.

Various federal, state, and local agencies collect mission-relevant information about chemical holdings. The West facility reportedly had not complied with all relevant and applicable regulatory requirements. For example, the facility reportedly had not contacted the Department of Homeland Security (DHS), which should have received information about any ammonium nitrate or anhydrous ammonia stored at the facility. The extent to which agencies shared relevant information about chemical holdings in order to enable effective regulatory oversight is still unresolved. As congressional policymakers consider the ramifications of the explosion in West, TX, they may face several policy issues. These policy issues include the:

- challenges arising from relying on reporting of chemical inventories by regulated facilities;
- potential for omission and duplication in existing regulatory reporting;
- long intervals between inspections at many such facilities;
- ability of federal, state, and local government agencies to share information effectively among themselves; and
- public and first-responder access to regulatory information.

INTRODUCTION

The explosion on April 17, 2013, at the West Fertilizer Company fertilizer distribution facility in West, TX, has led to questions about the oversight and regulation of agricultural fertilizer.[1] Congressional policymakers have expressed concern and requested information about coordination among federal regulatory agencies and the efficacy of agency regulations.[2] Different federal agencies apply a variety of regulatory structures to these chemicals. These regulations meet different policy rationales, goals, and purposes. Some attempt to ensure occupational safety, others environmental protection, and still others security. As a consequence, various federal, state, and local agencies collect mission-relevant information about chemical holdings, with some entities gaining a narrower, or broader, understanding of a facility's chemical inventories.

The West Fertilizer Company possessed a variety of agricultural chemicals at its retail facility,[3] but policy interest has focused on two chemicals: ammonium nitrate and anhydrous ammonia. In 2012, the West Fertilizer Company had reported to the U.S. Environmental ProtectionAgency (EPA) and the state of Texas the presence of 110,000 pounds of anhydrous ammonia and 270 tons of ammonium nitrate[4, 5] and had complied with certain provisions of the Clean Air Act,[6] but the West facility may not have complied with all reporting requirements. For example, the facility had not contacted the Department of Homeland Security (DHS), which should have received information about ammonium nitrate or anhydrous ammonia stored at the facility.[7] The DHS notes that the West Fertilizer Company facility might not have qualified as a regulated high-risk facility even if it had reported its chemical holdings.[8]

In addition to federal regulation requiring reporting and planning for ammonium nitrate and anhydrous ammonia, most state and some local governments have laws and regulations regarding the handling of either or both of these chemicals.[9] States regulate worker and environmental protection, product efficacy and safety, chemical sales and storage, and security. In addition, states may enforce standards such building, plumbing, mechanical, electrical, and fire codes. Local zoning laws may dictate placement of a facility within a community. Many states require registration to track the sellers and sales of fertilizer.[10]

This report will focus on some of the federal regulatory programs overseeing storage of ammonium nitrate and anhydrous ammonia by retailers. The report will not address federal regulation of material in transport.[11] It will discuss federal occupational safety, environmental, and security statutes and regulations applicable to each chemical. Select policy issues regarding these federal regulatory programs will be highlighted. It does not address various law enforcement activities related to tracking of anhydrous ammonia used for illegal drug synthesis (e.g., methamphetamine).

AMMONIUM NITRATE

Domestic production capacity of ammonium nitrate is approximately 7 million tons,[12] and it is produced worldwide. The vast majority of ammonium nitrate use occurs without incident. Most experts consider ammonium nitrate itself as a stable chemical with few handling restrictions, but, in combination with a fuel source, it can pose an explosion hazard.[13] Ammonium nitrate requires certain conditions, such as added heat or shock, confinement, or contamination to explode. Ammonium nitrate in combination with certain additives is widely used as a type of explosive known as a blasting agent. Ammonium nitrate is also used as a fertilizer.

Ammonium nitrate is suspected as the source of the explosion at the West Fertilizer Company.[14] Accidental explosions of ammonium nitrate have resulted in loss of lives and destruction of property. These accidents have rarely occurred, but have had high impacts. For example, the ammonium nitrate explosion in 1947 in Texas City, TX, where two ships carrying ammonium nitrate coated in wax and stored in paper bags caught fire and exploded, destroyed the entire dock area, including numerous oil tanks, dwellings, and business buildings.[15] Because of this history, and its inherent characteristics, a variety of federal agencies have authority to regulate ammonium nitrate, some of which has been delegated to state agencies.

Regulation by Bureau of Alcohol, Tobacco, Firearms and Explosives

The Bureau of Alcohol, Tobacco, Firearms and Explosives (ATF) of the Department of Justice (DOJ) regulates ammonium-nitrate-based blasting agents, but it does not regulate ammonium nitrate as fertilizer. TheATF has issued regulations regarding the necessary distance to be maintained between ammonium nitrate and other explosive materials.[16] The ATF and the fertilizer industry have undertaken joint voluntary efforts to secure ammonium nitrate fertilizer, predominantly through "know your customer" programs.[17]

Regulation by Department of Homeland Security

While often used for legitimate purposes, malicious actors might also use ammonium nitrate fertilizer as part of an improvised explosive device. For example, the bombing of the Murrah Federal Building in Oklahoma City, OK, used ammonium nitrate, along with other chemicals, as the primary explosive material. The ease with which ammonium nitrate fertilizer can be transformed into ammonium-nitrate-based blasting agent has led to increased scrutiny of its storage and transfer in order to prevent misappropriation and misuse.

For this reason, the Department of Homeland Security (DHS) has multiple statutory authorities under which it regulates ammonium nitrate for security purposes. In general, DHS authorities fall into two categories: securing facilities containing ammonium nitrate and tracking the transfer of ammonium nitrate.[18]

Since 2002, the U.S. Coast Guard (USCG) has regulated security at certain stationary facilities adjacent to waterways under the authority of the Maritime Transportation Security Act (MTSA, P.L. 107-295). In 2006, when Congress authorized DHS to regulate high-risk chemical facilities for security purposes in P.L. 109-295 (Department of Homeland Security Appropriations Act, 2007), it exempted those facilities already regulated under MTSA.[19] Congress also authorized DHS to oversee the sale and transfer of ammonium nitrate in P.L. 110-161 (Consolidated Appropriations Act, 2008).[20]

Maritime Transportation Security Act

The DHS, through the USCG, regulates security at facilities located adjacent to any waters subject to U.S. jurisdiction. Note that this regulatory requirement is irrespective of the presence of chemicals at the facility, but would capture any facility storing ammonium nitrate adjacent to waterways. The USCG implemented MTSA through a series of regulations issued in October 2003.[21] The MTSA requires the creation of an area security plan, assessment of security vulnerabilities at the facility, and the creation of a facility security plan. The USCG creates the area security plan. Either the USCG or the regulated facility performs a vulnerability assessment of the regulated facility. Based on this vulnerability assessment, the facility must develop and implement a security plan consistent with the broader USCG area maritime transportation security plan. The facility must update this plan every five years, and the USCG inspects the facility's implementation of the plan.[22]

Chemical Facility Anti-Terrorism Standards

In 2007, DHS issued an interim final rule establishing the Chemical Facility Anti-Terrorism Standards (CFATS), the regulatory program by which it implements its chemical facility security authority. Under the CFATS interim final rule, the Secretary of Homeland Security determines which chemical facilities must meet regulatory security requirements, based on the degree of risk posed by each facility. The DHS lists 322 "chemicals of interest" and screening threshold quantities for each chemical to determine the need to comply with CFATS.[23] The DHS considers each chemical in the context of three threats: release; theft or diversion; and sabotage and contamination. The regulation includes as chemicals of interest both pure ammonium nitrate and ammonium nitrate with more than 0.2% combustible substances, including any organic substance calculated as carbon, to the exclusion of any other added substance.[24] It identifies them as theft or diversion and release threats respectively.

The screening threshold quantity differs depending on the ammonium nitrate. Facilities possessing at least 5,000 pounds (400 pounds if packaged for transportation) of ammonium nitrate with more than 0.2% combustible substances are potentially high-risk facilities and must participate in the CFATS regulations.

Facilities possessing at least 2,000 pounds of transportable fertilizer mixture containing at least 33% ammonium nitrate are potentially high-risk facilities and must participate in the CFATS regulations. According to DHS, "This [screening threshold quantity] is geared toward ensuring that DHS secures [ammonium nitrate] inventories at major manufacturing and distribution facilities, as opposed to individual farmers involved mainly in the application of [ammonium nitrate]."[25]

Chemical facilities with greater than specified quantities of chemicals of interest must submit information to DHS. This information-gathering process is called the "Top-Screen." It enables DHS to determine each facility's risk status based on the information received. If a facility comes into possession of any of the chemicals of interest at the corresponding quantities, it must complete and submit a Top-Screen within 60 calendar days.[26] Also, the CFATS rule states that DHS may contact and instruct specific facilities to complete and submit Top-Screen information. Facilities that DHS deems high risk must meet CFATS requirements, which include performing a security vulnerability assessment, developing a site security plan, implementing the security plan, and then being inspected by DHS for compliance. See **Figure 1**.

Several types of facilities are exempt from the CFATS regulation. The statute exempts any facility that is: defined as a water system or wastewater treatment works; owned or operated by the Department of Defense or Department of Energy; regulated by the Nuclear Regulatory Commission (NRC); or regulated under MTSA. Many of these types of facilities comply with other security requirements.

The DHS has implemented a regulatory extension for agricultural chemical users, though not distributors or retailers.[27] Indeed, DHS stated in the CFATS interim final rule that "if a retail establishment does exceed any of these [screening threshold quantities], the retail establishment will have to complete the Top-Screen."[28] The CFATS rule does not, however, impose any limitations on the sale or transfer of ammonium nitrate.

Ammonium Nitrate Proposed Rule

In 2007, Congress enacted Section 563 of P.L. 110-161 (Secure Handling of Ammonium Nitrate), authorizing DHS to "regulate the sale and transfer of ammonium nitrate by an ammonium nitrate facility ... to prevent the misappropriation or use of ammonium nitrate in an act of terrorism."[29] The statute further required that DHS develop regulations wherein, among other responsibilities:

- sellers and purchasers of ammonium nitrate register with DHS and are screened by DHS against the Terrorist Screening Database (TSDB);
- sellers of ammonium nitrate verify purchaser's identify and registration, keep records of sales or transfers, and report theft or loss; and
- DHS audits, inspects, and monitors compliance; develops guidance materials; establishes an appeals process; and defines the threshold percentage of ammonium nitrate in a mixture for that mixture to be regulated.

The DHS began its ammonium nitrate rulemaking process with an advanced notice of proposed rulemaking in October 2008[30] and, in August 2011, issued a proposed rule regulating ammonium nitrate possession, sale, and transfer.[31]

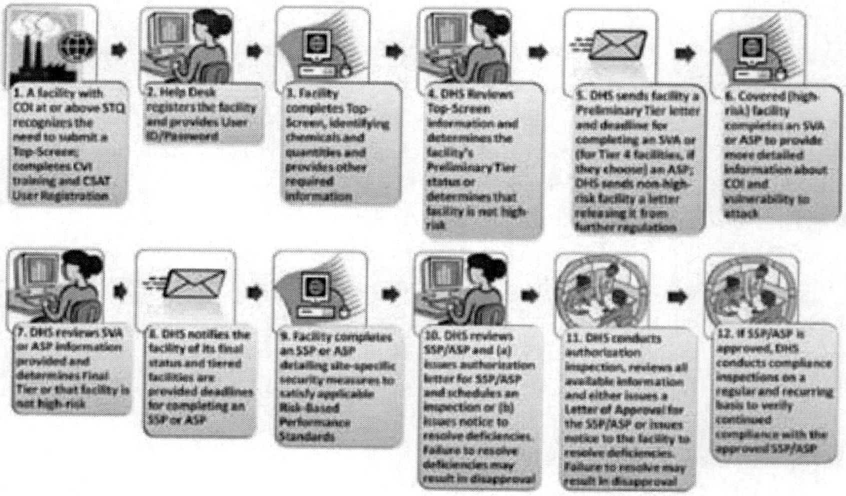

Source: Office of Infrastructure Protection, National Protection and Programs Directorate, Department of Homeland Security, *Chemical Facility Anti-Terrorism Standards (CFATS) and Ammonium Nitrate Security Regulation Update*, July 31, 2012.

Notes: COI = Chemical of Interest; STQ = Screening Threshold Quantity; CVI = Chemical-terrorism Vulnerability Information; CSAT = Chemical Security Assessment Tool; SVA = Security Vulnerability Assessment; ASP = Alternative Security Program; SSP = Site Security Plan.

Figure 1. Overview of CFATS Regulatory Process (July 2012).

The proposed rule would impose certain conditions on the sale or transfer of ammonium nitrate, for example, requiring that ammonium nitrate may only be transferred between registered ammonium nitrate sellers and registered ammonium nitrate purchasers. Each purchaser and seller would be required to apply to be registered with DHS, which would screen each applicant against the TSDB.[32] Following the screening process, approved individuals would be allowed to engage in the sale, purchase, or transfer of ammonium nitrate.

Transactions involving the sale or transfer of ammonium nitrate would be regulated at the point of sale and procedures for reporting a theft or loss of ammonium nitrate would be established. Ammonium nitrate sellers would be required to sell or transfer ammonium nitrate only to individuals who had successfully registered with DHS or those individuals' agents. The proposed rule would also require those selling ammonium nitrate to retain records for two years and report theft or loss of ammonium nitrate to federal authorities within 24 hours of discovery. The DHS would be responsible for ensuring compliance with these proposed standards to ensure public safety and support legitimate commerce.

This proposed rule received 132 comments, and DHS has not yet issued a final rule. According to DHS, it continues to adjudicate comments received on the proposed rule issued in August 2011.[33] Consequently, no regulation implements DHS's statutory authority to regulate ammonium nitrate sale and transfer.

Regulation by Occupational Safety and Health Administration

The Occupational Safety and Health Administration (OSHA) regulates employers within the agency's jurisdiction that use or possess certain quantities of ammonium nitrate at their worksites.[34] Under the "General Duty Clause" of the Occupational Safety and HealthAct (P.L. 91-596, as amended), each employer must provide its employees with a workplace that is free from "recognized hazards that are causing or are likely to cause death or serious physical harm" to its employees.[35] Thus, employers who use or possess ammonium nitrate must, regardless of any other regulations, ensure that the hazards known to be associated with ammonium nitrate do not put their employees in danger.

Nearly every workplace covered by OSHA is required to have an emergency action plan (EAP) that complies with OSHA Standard 1910.38.[36] An EAP must have, at minimum, the following elements:

- procedures for reporting a fire or other emergency; procedures for evacuation;
- procedures to be followed by employees who remain to operate parts of the facilities before evacuating;
- procedures to account for all employees after evacuation;
- procedures for employees performing rescue or medical duties; and
- names and job titles of persons who may be contacted by employees to provide information to employees about the EAP.[37]

According to information provided on its EPA Risk Management Plan (RMP), West Fertilizer Company reportedly had an EAP in place.[38]

In addition to their general duty and EAP requirements, employers regulated by OSHA must meet the following occupational safety and health standards that specifically apply to the use and possession of ammonium nitrate.[39]

Standard 1910.109: "Explosives and Blasting Agents"[40]

This standard regulates the storage, use, and transportation of explosives and blasting agents, including those that may contain ammonium nitrate. The general hazard provision of this standard states:

> No person shall store, handle, or transport explosives or blasting agents when such storage, handling, and transportation of explosives or blasting agents constitutes an undue hazard to life.[41]

In addition, this standard contains specific rules regarding the storage of ammonium nitrate.[42] These specific rules include height and spacing requirements for stored ammonium nitrate, required separation from other materials, and identification of certain building parameters. This standard applies to all persons storing 1,000 pounds or more of ammonium nitrate, as well as the owner or lessee of the structure containing the stored ammonium nitrate.[43]

Finally, the standard also specifies the minimum distances by which blasting agents and ammonium nitrate must be separated during storage.[44]

Standard 1910.1200: "Hazard Communication"[45]

This standard requires that chemicals used in the workplace be properly classified, labeled, and identified so that their potential safety and health hazards may be communicated to employees. For example, employers who use substances with recognized safety or health hazards are generally required to keep material safety data sheets on site and accessible to employees. In 2012, OSHA changed this standard to bring it more in line with the United Nations Globally Harmonized System of Classification and Labeling of Chemicals (GHS).[46] This standard generally applies to "any chemical which is known to be present in the workplace in such a manner that employees may be exposed under normal conditions of use or in a foreseeable emergency."[47]

Standard 1910.119: "Process Safety Management of Highly Hazardous Chemicals"

Section 304 of the Clean Air Act Amendments of 1990 (CAAA) mandated promulgation of the Process Safety Management (PSM) standard.[48] The PSM standard "contains requirements for preventing or minimizing the consequences of catastrophic releases of toxic, reactive, flammable, or explosive chemicals."[49] Processes involving ammonium nitrate are not subject to regulation under OSHA's process safety management of highly hazardous chemicals (PSM) standard, as ammonium nitrate is not listed as a highly hazardous chemical inAppendix A of the standard.[50]

Regulation by U.S. Environmental Protection Agency

The U.S. Environmental Protection Agency (EPA) is authorized to regulate production, distribution, storage, and release of most chemicals in commerce. Among the many environmental authorities the EPA relies upon, the Emergency Planning and Community Right-to-Know Act (EPCRA) and Section 112(r) of the Clean Air Act directly address the potential risks from facilities holding chemical hazards.

Emergency Planning and Community Right-to-Know Act

The accidental and sudden release of methyl isocyanate in an industrial accident at the Union Carbide Plant in December 1984 in Bhopal, India, and the attendant loss of life and injuries spurred legislative proposals to reduce the risk of chemical accidents in the United States. Enacted in 1986 as Title III of the Superfund Amendments and Reauthorization Act (SARA; P.L. 99-499), the Emergency Planning and Community Right-to-Know Act (EPCRA) established requirements and a framework to ensure that the EPA, state and local governments, and the private sector would work together to control and, if necessary, respond to releases of hazardous chemicals to the environment. Section 301 of EPCRA established state emergency response commissions (SERCs) and local emergency planning committees (LEPCs) to develop and implement local plans for coping with potential releases of hazardous chemicals.

Ammonium nitrate is regulated under two sections of EPCRA, Sections 311 and 312. EPCRA, Section 311, requires owners or operators of local facilities covered by the Occupational Safety and Health Act (P.L. 91-596, as amended) to submit a material safety data sheet (MSDS) for each "hazardous chemical," or a list of such chemicals, to the SERC, the LEPC, and the local fire department. An MSDS need only be submitted once, unless there is a significant change in the information it contains. An MSDS must be provided in response to a request by an LEPC or a member of the public. The definition of "hazardous chemicals" includes hundreds of chemicals.[51] EPCRA Section

311(e)(5) excludes certain substances from this definition, including "fertilizer held for sale by a retailer to the ultimate customer."[52]

EPCRA, Section 312, requires the same employers to submit annually an emergency and hazardous chemical inventory form to the SERC, LEPC, and local fire department. These forms must provide estimates of the maximum amount of the chemicals present at the facility at any time during the preceding year; estimates of the average daily amount of chemicals present; and the general location of the chemicals in the facility.[53] Information must be provided to the public in response to a written request. EPA is authorized to establish threshold quantities for chemicals below which facilities are not required to report.

Clean Air Act, Section 112(r)

In 1990, accumulated data on chemical accidents in the United States prompted Congress to amend Section 112 of the Clean Air Act (CAA) to further address the threat of catastrophic releases into the ambient air of chemicals that might cause immediate deaths or injuries in communities. Section 112(r)(7) requires owners and operators of covered facilities that produce, process, handle, or store certain extremely hazardous substances to submit to the EPA a risk management plan (RMP) for each process that might have an accidental chemical release. RMPs summarize the potential threat of unanticipated emissions into the ambient air of certain dangerous chemicals and facilities' plans to prevent such releases and mitigate any damage.

Ammonium nitrate is not a chemical regulated under Section 112(r) of the CleanAir Act. The implementing regulation applies to facilities that possess more than threshold levels of any of 77 acutely toxic substances or 63 flammable gases which the Administrator determined "pose the greatest risk of causing death, injury, or serious adverse effects to human health or the environment from accidental releases." In listing substances, the statute directed the EPA Administrator to consider "the severity of any acute adverse health effects associated with accidental releases of the substance," "the likelihood of accidental releases of the substance," and "the potential magnitude of human exposure to accidental releases of the substance." The law explicitly excludes fuels held for retail sale from regulation, as it does substances regulated under other sections of the Clean Air Act. In addition, the EPA Administrator was "authorized to establish a greater threshold quantity for, or to exempt entirely, any substance that is a nutrient used in agriculture when held by a farmer." The EPA Administrator did not include ammonium nitrate on the RMP chemical list.

ANHYDROUS AMMONIA

Ammonia has a variety of uses, for example, as an agricultural fertilizer and an industrial refrigerant. It also is used in industrial chemistry to manufacture other nitrogen-containing chemicals, such as ammonium nitrate, which is made from the reaction of ammonia with nitric acid. The term "anhydrous ammonia" refers to essentially pure ammonia that exists as a gas. This contrasts with "aqueous ammonia," where the ammonia is dissolved in water as a liquid solution. Many agricultural retailers and facilities with ammonia refrigeration systems store and use anhydrous ammonia. Domestic production of ammonia is approximately 12 million tons,[54] and it is produced worldwide.

In contrast to ammonium nitrate, where the concern is one of potential explosion or theft to illegally manufacture explosives, anhydrous ammonia is a toxic inhalational hazard. Generally, anhydrous ammonia is stored as a liquid under pressure; it becomes a toxic gas when released to the environment. Anhydrous ammonia can harm individuals who inhale or contact it. Thus, it is regulated to prevent release of the chemical into the atmosphere where it might travel as a cloud and impact workers and the surrounding environment. It is also encountered routinely in relatively small quantities by local law enforcement officials and hazardous material response teams as part of illegal drug synthesis operations (e.g., methamphetamine).

Regulation by Department of Homeland Security

The DHS uses the same authorities to regulate security of facilities possessing anhydrous ammonia that it uses to regulate those possessing ammonium nitrate: MTSA and CFATS. As mentioned above, in 2002, Congress enacted MTSA, which authorized the USCG to regulate security at facilities adjacent to waterways.[55] In 2006, Congress authorized DHS to regulate high-risk chemical facilities for security purposes, exempting those facilities already regulated under MTSA.[56]

Maritime Transportation Security Act

The MTSA required DHS, through the USCG, to regulate security at facilities located adjacent to any waters subject to U.S. jurisdiction. Note that this regulatory requirement is irrespective of the presence of chemicals at the facility, but would capture any facility storing anhydrous ammonia adjacent to waterways. The USCG implemented MTSA through a series of regulations issued in October 2003.[57] The MTSA requires the creation of an area security plan, assessment of security vulnerabilities at the facility, and the creation of a facility security plan. The USCG creates the area security plan. Either the USCG or the regulated facility performs a vulnerability assessment of the regulated facility. Based on this vulnerability assessment, the facility must develop and implement a security plan consistent with the broader USCG area maritime transportation security plan. The facility must update this plan every five years, and the USCG inspects the facility's implementation of the plan.[58]

Chemical Facility Anti-Terrorism Standards

In 2007, DHS issued an interim final rule establishing the Chemical Facility Anti-Terrorism Standards (CFATS), the regulatory program by which it implements its chemical facility security authority. The DHS regulates ammonia through the CFATS program. Under the CFATS interim final rule, the Secretary of Homeland Security determines which chemical facilities must meet regulatory security requirements, based on the degree of risk posed by each facility. The DHS lists 322 "chemicals of interest" and screening threshold quantities for each chemical to determine the need to comply with CFATS.[59] The regulation lists two formulations of ammonia (anhydrous ammonia and aqueous ammonia with a concentration of at least 20%) as chemicals of interest and identifies release threats for them.

Facilities holding 10,000 pounds or more of anhydrous ammonia or 20,000 pounds or more of aqueous ammonia with a concentration of at least 20% must submit information to DHS as part of the "Top-Screen" process. Facilities that DHS deems high risk based on the "Top-Screen" submission must meet CFATS requirements, which include performing a security vulnerability assessment, developing a site security plan, implementing the security plan, and then being inspected by DHS for compliance.

As noted above, the statute exempts several types of facilities from CFATS regulation. The DHS has implemented a regulatory extension for agricultural chemical users, though not distributors or retailers.[60] As DHS stated in the CFATS interim final rule, "if a retail establishment does exceed any of these [screening threshold quantities], the retail establishment will have to complete the Top-Screen."[61]

Regulation by Occupational Safety and Health Administration

OSHA regulates employers within its jurisdiction that use or possess certain quantities of anhydrous ammonia at their worksites. In addition to being subject to the General Duty Clause, employers regulated by OSHA must meet the following occupational safety and health standards that specifically apply to the use and possession of anhydrous ammonia.

Standard 1910.111: "Storage and Handling of Anhydrous Ammonia"[62]

This standard provides comprehensive requirements for the "design, construction, location, installation, and operation of anhydrous ammonia systems including refrigerated ammonia storage systems."[63] In addition to basic rules for the storage and handling of anhydrous ammonia and the operation of anhydrous ammonia systems, this standard provides additional requirements for different types of refrigerated and non-refrigerated storage systems and containers, transportation and transfer of anhydrous ammonia, and use of anhydrous ammonia in connection with farm vehicles such as spraying systems.

Standard 1910.119: "Process Safety Management of Highly Hazardous Chemicals"[64]

Section 304 of the Clean Air Act Amendments of 1990 (CAAA) mandated promulgation of the PSM standard.[65] The PSM standard "contains requirements for preventing or minimizing the consequences of catastrophic releases of toxic, reactive, flammable, or explosive chemicals."[66] The PSM standard covers processes involving anhydrous ammonia. As provided inAppendix A of the standard, OSHA considers anhydrous ammonia a highly hazardous chemical subject to standard when it is present in quantities exceeding 10,000 pounds.[67] Facilities with sufficient quantities of anhydrous ammonia must comply with the 14 elements of the process safety management regulation.

For the purposes of this standard, a process is defined as:

> any activity involving a highly hazardous chemical including any use, storage, manufacturing, handling, or the on-site movement of such chemicals, or combination of these activities. For purposes of this definition, any group of vessels which are interconnected and separate vessels which are located such that a highly hazardous chemical could be involved in a potential release shall be considered a single process.[68]

Any employer engaging in a process subject to the PSM standard must adhere to requirements designed to prevent an accidental release of a highly hazardous chemical or mitigate the consequences of such a release to prevent a catastrophic outcome. The CAAA requires 14 elements to be components of the PSM standard, and these elements generally have been grouped into the following requirements for employers:

1. develop and maintain written safety information identifying workplace chemical and process hazards, equipment used in the processes, and technology used in the processes;
2. perform a workplace hazard assessment, including, as appropriate, identification of potential sources of accidental releases, an identification of any previous release within the facility which had a likely potential for catastrophic consequences in the workplace, estimation of workplace effects of a range of releases, estimation of the health and safety effects of such range on employees;
3. consult with employees and their representatives on the development and conduct of hazard assessments and the development of chemical accident prevention plans and provide access to these and other records required under the standard;
4. establish a system to respond to the workplace hazard assessment findings, which shall address prevention, mitigation, and emergency responses;
5. periodically review the workplace hazard assessment and response system;
6. develop and implement written operating procedures for the chemical process including procedures for each operating phase, operating limitations, and safety and health considerations;
7. provide written safety and operating information to employees and train employees in operating procedures, emphasizing hazards and safe practices;
8. ensure contractors and contract employees are provided appropriate information and training;
9. train and educate employees and contractors in emergency response in a manner as comprehensive and effective as that required by the regulation promulgated pursuant to Section 126(d) of the Superfund Amendments and Reauthorization Act;
10. establish a quality assurance program to ensure that initial process related equipment, maintenance materials, and spare parts are fabricated and installed consistent with design specifications;
11. establish maintenance systems for critical process related equipment including written procedures, employee training, appropriate inspections, and testing of such equipment to ensure ongoing mechanical integrity;

12. conduct pre-start-up safety reviews of all newly installed or modified equipment;

13. establish and implement written procedures to manage change to process chemicals, technology, equipment and facilities; and

14. investigate every incident which results in or could have resulted in a major accident in the workplace, with any findings to be reviewed by operating personnel and modifications made if appropriate.[69]

The PSM standard does not apply to retail facilities. While the standard does not provide a definition of "retail facility," OSHA's interpretation is that the retail facility exemption applies to any establishment in which more than half of the income it obtains from the sale of highly hazardous chemicals covered by the PSM standard comes from direct sales to end users.[70] According to information provided on its EPA RMP, West Fertilizer Company reportedly was not covered by this standard.[71]

Standard 1910.1000: "Air Contaminants"[72]

This standard provides limits to employee exposure to certain substances. As provided in Table Z-1 of this standard, the permissible exposure limit (PEL) of an employee to anhydrous ammonia is:

- 50 parts of vapor or gas per million parts of contaminated air by volume at 25 degrees C and 760 Torr (50 ppm) as measured as an eight-hour time weighted average (TWA); or
- 35 milligrams of substance per cubic meter of air (35 mg/m^3, approximate) as measured as an eight-hour time TWA.[73]

Regulation by Environmental Protection Agency

The EPA is authorized to regulate production, distribution, storage, and release of most chemicals in commerce. Among the many environmental authorities the EPA relies upon, the Emergency Planning and Community Right-to-Know Act (EPCRA) and Section 112(r) of the Clean Air Act directly address the potential risks from facilities holding chemical hazards.

Emergency Planning and Community Right-to-Know Act

EPCRA Section 302 directs EPA to list "extremely hazardous substances" (EHS) that are to be the subject of emergency planning and to establish threshold planning quantities for each substance. Generally, EPA has included chemicals in this list if they can harm people exposed to them for only a short period of time. The implementing regulation requires owners or operators of facilities where an EHS is present at an amount above its threshold planning quantity to notify their SERC, which then is required to notify the EPA Administrator. Anhydrous ammonia is an EHS with a threshold planning quantity of 500 pounds.

EPCRA Section 303 requires LEPCs to work with facilities handling EHS to develop response procedures, evacuation plans, and training programs for people who will be the first to respond in the event of an accident. Upon request, facility owners and operators are required to provide an LEPC with any additional information that it finds necessary to develop or implement an emergency plan.

EPCRA Section 304 requires that facilities immediately report a release of any EHS or any "hazardous substance" (a much broader category of chemicals defined under the Comprehensive Environmental Response, Compensation, and Liability Act (CERCLA) Section 102(a)) that exceeds the reportable quantity to appropriate state, local, and federal officials.[74] Releases of a reportable quantity of a "hazardous substance" also must be reported to the National Response Center under CERCLA Section 103(a).[75] (For more on CERCLA, see CRS Report RL30798, *Environmental Laws: Summaries of Major Statutes Administered by the Environmental Protection Agency*, which includes a summary of CERCLA.) EPCRA Section 304 also would require reporting of any release above the reporting threshold for anhydrous ammonia.

As noted in the section above for ammonium nitrate, EPCRA Section 311 requires owners or operators of local facilities regulated by OSHA to submit an MSDS for each "hazardous chemical," or a list of such chemicals, to the LEPC, the SERC, and the local fire department. The OSHA defines "hazardous chemicals," but EPCRA Section 311(e)(5) excludes "fertilizer held for sale by a retailer to the ultimate customer" from this definition.

Also as described previously, EPCRA Section 312 requires the same employers to submit annually an emergency and hazardous chemical inventory form to the LEPC, SERC, and local fire department. West Fertilizer Company reported their chemical inventory to the state of Texas.[76]

Clean Air Act, Section 112(r)

As described above, the CAA Section 112(r)(7) requires owners and operators of covered facilities that produce, process, handle, or store certain extremely hazardous substances, including, but not necessarily limited to extremely hazardous substances under EPCRA, to submit Risk Management

Plans (RMPs) to the EPA. RMPs summarize the potential threat of unanticipated emissions into the ambient air of certain dangerous chemicals and facilities' plans to prevent such releases and mitigate any damage. Congress explicitly required planning for possible releases of anhydrous ammonia.

POLICY ISSUES

As congressional policymakers consider the ramifications of the explosion in West, TX, they may face several policy issues. These policy issues include the challenges arising from relying on reporting of chemical inventories by regulated facilities; the potential for omission and duplication in existing regulatory reporting; the long intervals between inspection at many such facilities; the ability of federal, state, and local government agencies to share information effectively among themselves; and public and first-responder access to regulatory information.

Challenges from Reporting by Facilities

Congressional policymakers might consider mechanisms to address the data quality issues arising from reporting by facilities. Regulatory programs often require facilities to report their chemical holdings, but such reporting creates the potential for the regulatory agency to be unaware if the facility does not report. This may be especially true for federal agencies that rely on state or local implementation of federal requirements. In addition, the quality of such information may be questioned due to its self-reported nature.[77] For example, the Government Accountability Office has found that DHS does not use self-reported vulnerability assessment information in determining the security risk at chemical facilities.[78]

Lack of Harmonized Reporting

Because different chemicals often have different reporting requirements, regulated entities often report different information to different regulating agencies. For example, as noted above, DHS, EPA, and state regulatory agencies receive different information; each chemical may fall under some, all, or no regulatory requirements due to particular exclusions or exemptions. Consequently, no central overview of facility chemical holdings exists, and may create the potential for inefficient regulation.

Inspection Rate

Interested Members of Congress might focus upon the rate of inspection under the various regulatory programs. Under some federal programs, inspection has not occurred at a pace expected by policymakers. For example, the DHS stated that it had inspected and approved site security plan implementation at 166 of 4,298 facilities regulated under CFATS as of July 2013.[79] In 2009, the EPA Office of Inspector General reported that EPA had inspected less than half of the EPA-identified high-risk facilities and less than 35% of all facilities reporting under Section 112(r) of the Clean Air Act.[80] Inspection of facilities reporting their chemical holdings may increase the trustworthiness of supplied information. Also, regular inspection may yield benefits in enhanced compliance with best practices and regulatory adherence. Of course, increased inspection relies upon a substantial inspector cadre, which would require additional appropriations for support. Similarly, delegation of inspection responsibilities for federal programs to state agencies might reduce the number of discrete inspections but increase the costs borne by those states. Conversely, additional efficiency might be achieved with a cross-trained inspectorate that might inspect a facility under all, or most, applicable regulatory regimes.

Ability of Government Agencies to Share Information Effectively

Congressional policymakers have identified information sharing in the homeland security and intelligence context as a high priority following the events of September 11, 2001. The reported gap in information between EPA, which knew of the West Fertilizer Company's existence and its anhydrous ammonia holdings, and DHS, which did not, indicates a potential gap in information sharing among federal regulators. Similarly, information possessed by the state of Texas, which knew of the West Fertilizer Company's ammonium nitrate holdings, apparently did not reach DHS. The DHS has detailed some of its information sharing activities, including sharing CFATS-regulated facility information with federal, state, and local partners, and is reinvigorating its coordination with EPA and review of EPA RMP data to identify facilities that may have failed to submit information under CFATS.[81] Increased information sharing may also raise concern among industry stakeholders regarding the potential for inadvertent release of proprietary information due to the effects of aggregation and the increased need to transmit such information between different entities. President Obama issued an executive order on improving chemical facility safety and security, which would aim to enhance federal agency coordination and information sharing, among other activities.[82]

Public Access to Information

Congressional policymakers have long been interested in the balance between providing public access to information regarding chemical hazards and limiting the potential for malicious use of such information.[83] Reporting under EPA regulation is intended to improve accident planning through increasing community and first-responder awareness of the potential for release of hazardous materials. First responders receive some of this information directly and residents of the surrounding community may obtain access to it. In contrast, DHS restricts information dissemination regarding security efforts at chemical facilities, partly in order to limit the ability of malicious actors to use such information for targeting purposes. The extent to which the residents and first responders of West, TX, knew of the potential hazards may provide insights into this balance and the effectiveness of community outreach efforts.

End Notes

[1] For example, see Manny Fernandez and Steven Greenhouse, "Texas Fertilizer Plant Fell through Regulatory Cracks," *New York Times*, April 24, 2013.

[2] For example, see Representative Michael T. McCaul, Chairman, Committee on Homeland Security; Representative Fred Upton, Chairman, Committee on Energy and Commerce; and Representative John Carter, Chairman, Homeland Security Appropriations Subcommittee, *Letter to Janet Napolitano, Secretary, U.S. Department of Homeland Security*, July 22, 2013; Senator Tom Carper, Chairman, Committee on Homeland Security and Governmental Affairs, *Letter to Janet Napolitano, Secretary, and Suzanne Spaulding, Acting Under Secretary, National Protection and Programs, U.S. Department of Homeland Security*, June 28, 2013; Representative Henry A. Waxman, Ranking Member, Committee on Energy and Commerce, and Representative Bennie G. Thompson, Ranking Member, Committee on

Homeland Security, *Letter to President Barack Obama*, May 2, 2013; Senator Barbara Boxer, Chairman, Committee on Environment and Public Works, *Letter to Bob Perciasepe, Acting Administrator, U.S. Environmental Protection Agency*, April 30, 2013; and Representative George Miller, Senior Democratic Member, Committee on Education and the Workforce, and Representative Joe Courtney, Senior Democratic Member, Subcommittee on Workforce Protections, Committee on Education and the Workforce, *Letter to Gene L. Dodaro, Comptroller General, U.S. Government Accountability Office*, April 25, 2013.

[3] Adair Grain, Inc. DBA West Fertilizer Co., Tier II, Emergency and Hazardous Chemical Inventory, 2012, http://op.bna.com/env.nsf/id/jsun-96vvcm/$File/Adair%20Grain%20Inc%202012%20Tier%202%20Report.pdf.

[4] It is unclear from the Tier II report whether this reflects maximum amounts stored on site or some other value. Investigators have estimated that approximately 150 tons of ammonium nitrate were present on the day of the explosion (Randy Lee Loftis, "Analysis: West Fertilizer Report Details Sequence of a Catastrophe," *Dallas Morning News*, May 16, 2013).

[5] Reportedly, 28 to 34 tons of ammonium nitrate, only a portion of the amount present, exploded (Randy Lee Loftis, "Analysis: West Fertilizer Report Details Sequence of a Catastrophe," *Dallas Morning News*, May 16, 2013).

[6] Right-to-Know Net, West Fertilizer Co. Risk Management Plan, June 30, 2011, http://data.rtknet.org/rmp/rmp.php?facility_id=100000135597&database=rmp&detail=3&datype=T.

[7] Personal communication between DHS and CRS staff, April 23, 2013.

[8] Department of Homeland Security, *DHS Responses to Rep. McCaul and Rep. Meehan's May 2, 2013 Letter Regarding the Chemical Facility Anti-Terrorism Standards (CFATS) Program*, June 2013.

[9] The authors thank Jerry Yen, Intern, Congressional Research Service, for research assistance on state fertilizer regulation.

[10] The Association of American Plant Food Control Officials and the Fertilizer Institute have compiled a list of state fertilizer laws and regulations. Association of American Plant Food Control Officials, "State Fertilizer Laws and Regulations," at http://www.aapfco.org/state_laws_regs.html and The Fertilizer Institute, "2012 Summary of State Fertilizer Laws," at http://www.tfi.org/node/507.

[11] The U.S. Department of Transportation and the U.S. Coast Guard regulate the transportation of chemicals by land and sea.

[12] IFDC, *North America Fertilizer Capacity*, March 2013.

[13] Environmental Protection Agency, Office of Solid Waste and Emergency Response, Chemical Emergency Preparedness and Prevention Office, "Chemical Safety Alert: Explosion Hazard from Ammonium Nitrate," EPA 550-F-97-002d, December 1997.

[14] No final determination of the root cause of the explosion of the West Fertilizer Company has been made. The Chemical Safety Board has begun an investigation into the explosion. Chemical Safety Board, "Chemical Safety Board Deploying to West Fertilizer Plant Accident," *Press Release*, undated.

[15] S. Olafson, "Texas City Just Blew Up," *Houston Chronicle*, April 16, 1997.

[16] 27 C.F.R. §550.220.

[17] See, for example, http://www.atf.gov/explosives/programs/ammonium-nitrate-security/index.html.

[18] The U.S. Coast Guard also regulates transport of ammonium nitrate on waterways as a "certain dangerous cargo."

[19] Section 550, P.L. 109-295, Department of Homeland Security Appropriations Act, 2007.

[20] Section 563, P.L. 110-161, Consolidated Appropriations Act, 2008.

[21] 33 C.F.R. §105.

[22] Vessels have similar requirements to facilities under the MTSA.

[23] 72 *Federal Register* 65396-65435 (November 20, 2007).

[24] Note that DHS does not regulate combinations of ammonium nitrate/fuel oil (ANFO) under the CFATS program (Department of Homeland Security, *CFATS Knowledge Center FAQ Number 1383*, May 22, 2009).

[25] 72 *Federal Register* 65396-65435 (November 20, 2007) at 65407.

[26] 72 *Federal Register* 17688-17745 (April 9, 2007) at 17731.

[27] See Section 550(a), P.L. 109-295, Department of Homeland Security Appropriations Act, 2007, and 73 *Federal Register* 1640 (January 9, 2008).

[28] 72 *Federal Register* 17688-17745 (April 9, 2007) at 17697.

[29] Section 563, P.L. 110-161, Consolidated Appropriations Act, 2008.

[30] 73 *Federal Register* 64280-64282 (October 29, 2008).

[31] 76 *Federal Register* 46907-46957 (August 3, 2011).

[32] The Terrorist Screening Database (TSDB) is a centralized federal database of information about known or suspected terrorists. For more information, see http://www.fbi.gov/about-us/nsb/tsc/tsc_faqs.

[33] Department of Homeland Security, *DHS Responses to Rep. McCaul and Rep. Meehan's May 2, 2013 Letter Regarding the Chemical Facility Anti-Terrorism Standards (CFATS) Program*, June 2013.

[34] OSHA does not have jurisdiction over public sector employers; employers who are regulated by other federal safety agencies, such as the Mine Safety and Health Administration (MSHA); and employers in states with approved occupational safety and health plans. Currently, 20 states and Puerto Rico have their own occupational safety and health plans that cover the private sector. Under Section 18(c) of the Occupational Safety and Health Act [29 U.S.C. §667(c)], a state plan must contain regulations "at least as effective" as those enforced by OSHA.

[35] 29 U.S.C. §654(a).

[36] 29 C.F.R. §1910.38.

[37] 29 C.F.R. §1910.38(c).

[38] Right-to-Know Net, Risk Management Plan, http://data.rtknet.org/rmp/rmp.php?facility_id =100000135597& database=rmp&detail=3&datype=T.

[39] Pursuant to Section 6 of the Occupational Safety and Health Act (29 U.S.C. §655) OSHA may promulgate occupational safety and health standards through regulation.

[40] 29 C.F.R. §1910.109.

[41] 29 C.F.R. §1910.109(b)(1).

[42] 29 C.F.R. §1910.109(i).

[43] 29 C.F.R. §1910.109(i)(2)(i).

[44] 29 C.F.R. §1910.109(g)(4)(vi).

[45] 29 C.F.R. §1910.1200.

[46] Department of Labor, Occupational Safety and Health Administration, "Hazard Communication: Final Rule," 77 *Federal Register* 17574, (March 26, 2012).

[47] 29 C.F.R. §1910.1200(b)(2).

[48] P.L. 101-549.

[49] 29 C.F.R. §1910.119.

[50] 29 C.F.R. §1910.119. A similar PSM standard for the construction industry is at 29 C.F.R. §1926.64.

[51] 29 C.F.R. 1910.1200(c) defines hazardous chemical to mean "any substance, or mixture of substances" that "is classified as a physical hazard or a health hazard, a simple asphyxiant, combustible dust, pyrophoric gas, or hazard not otherwise classified." "Health hazard means a chemical which is classified as posing one of the following hazardous effects: acute toxicity (any route of exposure); skin corrosion or irritation; serious eye damage or eye irritation; respiratory or skin sensitization; germ cell mutagenicity; carcinogenicity; reproductive toxicity; specific target organ toxicity (single or repeated exposure); or aspiration hazard." "Physical hazard means a chemical that is classified as posing one of the

following hazardous effects: explosive; flammable (gases, aerosols, liquids, or solids); oxidizer (liquid, solid or gas); self-reactive; pyrophoric (liquid or solid); self-heating; organic peroxide; corrosive to metal; gas under pressure; or in contact with water emits flammable gas."

[52] EPCRA Section 311(e) excepts foods, food additives, or other substances regulated by the Food and Drug Administration; solids in a manufactured item to the extent exposure to people or the environment does not occur; substances used for personal or household purposes; substances used in research or hospitals; and substances used in routine agricultural operations.

[53] EPCRA allows facilities to report aggregate amounts of chemicals with similar health and environmental effects. This is called "Tier I" information. However, chemical-specific information ("Tier II") must be provided on request (under certain conditions) to an SERC, LEPC, fire department, or the public.

[54] IFDC, *North America Fertilizer Capacity*, March 2013.

[55] The MTSA also provided additional authorities, including regulating security on vessels.

[56] Section 550, P.L. 109-295, Department of Homeland Security Appropriations Act, 2007.

[57] 33 C.F.R. §105.

[58] Vessels have similar requirements to facilities under the MTSA.

[59] 72 *Federal Register* 65396-65435 (November 20, 2007).

[60] See Section 550(a), P.L. 109-295, Department of Homeland Security Appropriations Act, 2007; and 73 *Federal Register* 1640 (January 9, 2008).

[61] 72 *Federal Register* 17688-17745 (April 9, 2007) at 17697.

[62] 29 C.F.R. §1910.111.

[63] 29 C.F.R. §1910.111(a)(1)(i).

[64] 29 C.F.R. §1910.119.

[65] P.L. 101-549.

[66] 29 C.F.R. §1910.119.

[67] 29 C.F.R. §1910.119 Appendix A.

[68] 29 C.F.R. §1910.119(b).

[69] 29 U.S.C. §655 note.

[70] See for example OSHA's standard interpretation "Clarification of the retail facilities exemption under the PSM standard," December 12, 2005, https://www.osha.gov/pls/oshaweb/owadisp.show_document?p_table= INTERPRETATIONS&p_id=25171.

[71] Right-to-Know Net, Risk Management Plan, http://data.rtknet.org/rmp/rmp.php?facility_id =100000135597& database=rmp&detail=3&datype=T.

[72] 29 C.F.R. §1910.1000.

[73] 29 C.F.R. §1910.1000 Table Z-1.

[74] Under CERCLA Section 102(a), a "hazardous substance" includes any "elements, compounds, mixtures, solutions, and substances which, when released into the environment may present a substantial danger to the public health or welfare or the environment." Included in this definition are substances listed under the authority of any of the major environmental statutes (see CERCLA Section 101(14)).

[75] The term "release" is defined in CERCLA Section 101(22) to exclude a release which results in exposure to persons solely within a workplace and normal application of fertilizer.

[76] Adair Grain, Inc. DBA West Fertilizer Co., Tier II, Emergency and Hazardous Chemical Inventory, 2012, http://op.bna.com/env.nsf/id/jsun-96vvcm/$File/Adair%20Grain%20Inc%202012%20Tier%202%20Report.pdf. It is unclear whether the amounts reported are maximum amounts stored on site or some other value.

[77] The Tier II reporting document filed by West Fertilizer Company, for example, has several internal inconsistencies regarding its chemical inventory.

[78] Government Accountability Office, *Critical Infrastructure Protection: DHS Efforts to Assess Chemical Security Risk and Gather Feedback on Facility Outreach Can Be Strengthened*, GAO-13-353, April 2013.

[79] Department of Homeland Security, *Chemical Facility Anti-Terrorism Standards*, July 2013, http://www.dhs.gov/sites/default/files/publications/CFATS%20Media%20FS_July2013.pdf.

[80] Office of Inspector General, Environmental Protection Agency, *EPA Can Improve Implementation of the Risk Management Program for Airborne Chemical Releases*, Report No. 09-P-0092, February 10, 2009, p. 15. See also Office of Inspector General, Environmental Protection Agency, *Improvements Needed in EPA Training and Oversight for Risk Management Program Inspections*, Report No. 13-P-0178, March 21, 2013.

[81] Department of Homeland Security, *DHS Responses to Rep. McCaul and Rep. Meehan's May 2, 2013 Letter Regarding the Chemical Facility Anti-Terrorism Standards (CFATS) Program*, June 2013.

[82] Office of Press Secretary, The White House, *FACT SHEET: Executive Order on Improving Chemical Facility Safety and Security*, August 1, 2013.

[83] For example, Congress addressed this issue in the context of EPA plans to electronically publish potential consequence data from RMP submissions by barring such publication (Section 3 of P.L. 106-40, Chemical Safety Information, Site Security and Fuels Regulatory Relief Act). See also House Committee on Commerce, Subcommittee on Health and Environment and Subcommittee on Oversight and Investigations, "Internet Posting of Chemical 'Worst Case' Scenarios: A Roadmap for Terrorists," *Serial No. 106-3*, February 10, 1999.

In: Ammonium Nitrate Facilities
Editor: Allegra Ward

ISBN: 978-1-63321-716-4

Chapter 3

CHEMICAL ADVISORY: SAFE STORAGE, HANDLING, AND MANAGEMENT OF AMMONIUM NITRATE*

Environmental Protection Agency, Occupational Safety and Health Administration, The Bureau of Alcohol, Tobacco, Firearms and Explosives

The Environmental Protection Agency (EPA), the Occupational Safety and Health Administration (OSHA), and the Bureau of Alcohol, Tobacco, Firearms, and Explosives (ATF) ("we") are issuing this advisory[1] as part of an ongoing federal effort to improve chemical risk management, and to advance safety and protect human health and the environment. This advisory contains information on recent and past accidents involving AMMONIUM NITRATE (commonly referred to as AN), on the hazards of AN, how to manage these hazards, and appropriate steps for community emergency planning and proper emergency response. It is focused primarily on safe handling and storage of higher density, solid AN pellets and prills (a prill is a small bead) used in fertilizers. This advisory is intended to broadly disseminate lessons learned

* This is an edited, reformatted and augmented version of a document, EPA 550-S-13-001, issued August 2013.

from recent incidents involving AN so that such incidents can be prevented in the future. Also provided is a list of information resources, including relevant codes and standards, industry publications, and applicable statutes and regulations that will help facilities handling AN and first responders better understand the hazards so they can effectively manage the risks. The information provided is not intended to cover all the hazards, safe practices or technical challenges associated with the manufacturing of AN; liquid fertilizers containing AN; manufacturing, storage or use of explosives or blasting agents containing AN; or the transportation of AN. For these particular situations, please consult other sources including the appropriate references, standards and regulations, cited at the end of this document.

ACCIDENTS

In general, AN is manufactured for use as a fertilizer and to produce explosives and blasting agents.[2] There are several other uses in the chemical industry, such as the production of nitrous oxide. These other uses represent a small fraction of amount of AN used in the US.

Although pure AN is stable at ambient temperature and pressure under many conditions, the chemical itself does not burn. AN is a strong oxidizer[3] and it supports and accelerates the combustion of organic (and some inorganic) material, increasing the fire hazard and complicating the fire fighting challenges. AN may explode when exposed to strong shock or when subjected to high temperatures in confinement.

Millions of tons of AN are produced annually in the US. Incidents involving AN are rare, but as is shown in the accidents below, they can have severe consequences. Most recently, on April 17, 2013, a fire at a fertilizer storage and distribution facility in West, Texas, resulted in a detonation of AN fertilizer stored at the facility, killing 15 people, including some of the firefighters responding to the fire. That incident remains under investigation,[4] but much has been learned from other AN explosions.

- On October 2, 2003, a fire and explosion occurred in a double story farm warehouse in St. Romain en Jarez, France, involving 3 to 5 tons of AN stored in bags. This incident killed 26 people, 18 of whom were firefighters. In this incident, improper storage methods are thought to have played a role.

- On September 21, 2001, a massive explosion occurred in a warehouse at the Azote de France fertilizer factory in Toulouse, France, involving 200-300 tons of AN, which was stored in bulk in a hangar. The explosion resulted in the death of 30 people, 2500 injuries, the destruction of the factory, and an additional 10,000 buildings being heavily damaged. The exact cause of this accident remains unknown. Storage of incompatible material with AN is believed to have been a factor.

We have learned several key lessons as a result of these accidents and additions studies of AN, including:

– The conditions of storage and the materials co-located with AN while in storage are crucial to the safety and stability of the AN.

Explosions of stored AN are responsible for some of the worst chemical disasters on record. Several of these incidents, including two in Germany in 1921, occurred during attempts to break up large piles of solidified or caked AN and ammonium sulfate mixtures using explosives. In both cases, the initial blast intended to break up solid AN initiated an unintended general detonation of the AN or ammonium sulfate mixture.

– AN will self-confine under some conditions. Adding heat, such as a booster charge intended to break up clumps, can initiate a general detonation of the AN.

Other large explosions have been triggered by fires involving AN in confined spaces, including the 1947 explosion in Texas City, Texas, of two cargo ships. In that case, the first ship is thought to have exploded due to a fire in the hold involving AN fertilizer that had been manufactured with a wax coating and stored in paper bags. The wax would have been one potential source of fuel for mixing with the AN, thus creating an explosive situation. The second ship exploded some time later, likely due to a fire caused by the first explosion. These two explosions resulted in deaths of nearly 600, including all but one member of the Texas City Fire Department.

As a result of such accidents and subsequent studies of the properties of AN, caked AN is no longer broken up with explosive materials, and organic material such as wax coatings are no longer used for AN fertilizer.

Our intent in issuing this advisory is to identify actions that should be taken as a result of the lessons learned from the more recent accidents involving AN. Similar to the corrective steps taken following the 1921 and 1947 incidents, this advisory emphasizes the safe steps that should become common practice in the industry and emergency response community in order to prevent the catastrophic loss of life and property damage.

Here are some of the things we have learned from accidents involving AN:

> AN will self-compress/self-confine under some conditions, becoming much more likely to explode.
>
> AN is at risk for explosion when stored near other material that can add fuel to the AN – such as grain, sugar, seeds, sawdust, and most especially petroleum fuels such as diesel.
>
> AN is a powerful oxidizer and a rich source of nitrate, which provides energy to an explosion. Thus, the presence of fuel and/or heat (and especially both) near AN is a very high hazard situation.

INFORMATION ON HAZARDS

Hazard Classification

For the purpose of transportation, AN that contains less than 0.2 percent combustible substances and AN fertilizers are classified by the U.S. Department of Transportation (DOT), as oxidizers. *AN with more than 0.2 percent combustible substances is classified by DOT as an explosive.*[5] (see box below).

The National Fire Protection Association (NFPA) assigns an instability rating of 3 (in a range of 0-4) to AN, meaning AN is capable of detonation, explosive decomposition, or explosive reaction, but that a strong initiating source or confinement in extreme temperatures is required. AN can explode under certain conditions by adding energy (heat, shock), especially when contaminants are present or it is under confinement.

> "Pure" ammonium nitrate is stable and will explode only under extraordinary circumstances.

However, the addition of combustible materials such as sugar, grain dust, seed husks or other organic contaminants, even in fairly low percentages, creates a dangerous combination and the ammonium nitrate mixture becomes far more susceptible to detonation. This characteristic of ammonium nitrate underlies most of the advice and recommendations for safe handling contained herein.

Decomposition Chemistry

AN melts at 337° F (170° C) and begins to undergo decomposition when molten. Hazardous scenarios with AN can involve simple thermal decomposition initiated by external fire or other heating, self-sustained decomposition also known as "cigar burning," and detonation.

Decomposition creates toxic gases containing ammonia and nitrogen oxides. The resulting nitrogen oxides will support combustion, even in the absence of other oxygen. The resulting heat and pressure from the decomposition of AN may build up if the reaction takes place in a confined space and the heat and gases created are not able to dissipate. As the temperature rises, the rate of decomposition increases. In a confined space, the pressure can reach dangerous levels and cause an explosion that will include the detonation of the AN.

When dealing with a large quantity of AN, localized areas of high temperature may be sufficiently confined by the mass of material to initiate an explosion. The explosion of a small quantity of AN in a confined space (e.g., a pipe) may act as a booster charge and initiate the explosion of larger quantities (e.g., in an associated vessel).

During a fire in a facility where AN is present, the AN can become hot and molten which makes the material very sensitive to shock and detonation, particularly if it becomes contaminated with incompatible material such as combustibles, flammable liquids, acids, chlorates, chlorides, sulfur, metals, charcoal, sawdust, etc. If a molten mass becomes confined (e.g., in drains, pipes or machinery), it can explode.

Most types of AN do not continue to decompose once a fire has been extinguished. However, some types of AN fertilizers containing a small percentage of chlorides (e.g., potassium chloride) undergo a smoldering (self-sustaining) decomposition that can spread throughout the mass to produce substantial toxic fumes, even when the initial heat source is removed. These fertilizers that can self-sustain decomposition, known as "cigar burners" are normally compound fertilizers that contain between 5% to 25% nitrogen from ammonium nitrate, up to 20% phosphate (as P_2O_5) and chloride (which may only be present as a small percentage).

Contaminants

AN mixed with oil or other sensitizing contaminants may explode or detonate when exposed to fire or shock. Organic materials (e.g., packing materials, seed, etc.) will increase the likelihood of an explosion and will make the AN explosion more energetic.

AN may also be sensitized by certain inorganic contaminants, including chlorides and some metals, such as aluminum powder, chromium, copper, cobalt, and nickel.

As AN solution becomes more acidic, its stability decreases, and it may be more likely to explode.

Solid AN readily absorbs moisture, which can lead to caking, self-compression and self confinement. This in turn increases susceptibility to explosion in a fire.

The density, particle size and concentration of solid AN in a material, as well as the presence of other additives, affects the hazard of the material. The technical grade of AN is a lower density (higher porosity) prilled material. Higher density prills are used as fertilizer. AN can be fused with ammonium sulfate fertilizer or amended with carbonate materials to reduce its explosive properties. More information on additives is discussed in *Guidance for the Storage, Handling and Transportation of Solid Mineral Fertilizers* found in the Reference section. Solid fertilizers are usually coated with an inorganic, non-combustible anti-caking compound to prevent sticking and clumping.

AN in undiluted or pure form has a higher degree of overall hazard than when it is mixed or blended with compatible or non-combustible materials that can reduce the concentration. In general for fertilizer blends containing AN, the more nitrogen they contain, the greater the explosion hazard they pose. Blended fertilizers containing AN and chloride compounds and blended fertilizers containing AN contaminated with combustible materials or incompatible substances pose increased explosion hazards. A large number of blended fertilizers are produced from basic primary fertilizer products (e.g., ammonium nitrate, urea, and mono-ammonium phosphate) and natural materials (e.g., rock phosphate, potassium chloride) which can introduce contaminants. All such materials are not necessarily compatible with each other and some may produce undesirable effects when mixed with others. These undesirable effects can include, for example, chemical reaction(s) and physical effects (e.g. stickiness which can cause handling difficulties, moisture migration giving rise to caking tendency). Facilities can consult *Guidance for Compatibility of Fertilizer Blending Materials* listed in the Reference section to assess potential incompatibility. The Safety Data Sheet (SDS – formerly MSDS) of the AN product should be used as one source of information to assess the overall hazard. The effects of added components can only be determined after careful review of the SDS and other available hazard literature.

> Confinement and/or the addition of fuel to AN creates a real danger of explosion. The addition of heat when either of these conditions exists can lead to disaster. Accordingly, the responder should quickly assess if AN has been involved in the fire and whether the AN has been compromised in any of these ways, and plan the fire response accordingly.

HAZARD REDUCTION

What Steps Should Facility Owners or Operators Take to Reduce the Hazards of AN During Storage And Handling?

Storage/Process Conditions to Avoid

Persons engaged in the handling, management or emergency planning for AN must be aware of the hazards of AN and ensure that the conditions that may lead to an explosion are not present. Measures that facilities should take to ensure the safe storage, use and handling of AN include:

- Avoid heating AN in a confined space.
  - Processes involving AN should be designed to avoid this possibility.
  - Avoid localized heating of AN, potentially leading to development of high temperature areas (e.g., AN fertilizer should not be stored near sources of heat such as steam pipes, radiators, hot ducts, light bulbs etc.).
- Ensure that AN is not exposed to strong shock waves from explosives. AN storage near high explosives or blasting agents must conform to

ATF's Table of Separation Distances, Title 22 of the Code of Federal Regulations, section 555.220 (22 CFR 555.220).

- Avoid contamination of AN with combustible materials or organic substances such as packing materials, dust, seed, oils, and waxes.
  - If possible, do not co-locate AN, especially bulk AN in bins, with dust-producing organics such as grains or seeds.
- Avoid contamination of AN with inorganic materials that may contribute to its sensitivity to explosion, including chlorides and some metals, such as aluminum powder, chromium, copper, cobalt, and nickel.
  - Pay attention to the materials used to build storage areas and cribs. Wood and aluminum or other metals must be specially treated to prevent impregnation if they are going to be in contact with AN. Metal materials can be treated with epoxy tar or chlorinated rubbers to prevent corrosion of the metal and contamination of the AN.
- Maintain the pH of AN solutions within the safe operating range of the process. In particular, avoid low pH (acidic) conditions.
  - If possible, do not co-locate acids in an AN storage area.
- Keep molten or solid AN out of confined spaces, especially sewers or drains where it can react with organic materials there.

Certain specific safety and handling instructions (required and recommended) apply for safe handling and storage of AN[6] under certain conditions:

OSHA's standard for Explosives and Blasting Agents at 29 CFR 1910.109(i) contains requirements for AN stored in the form of crystals, flakes, grains or prills including fertilizer grade, dynamite grade, nitrous oxide grade, technical grade, and other mixtures containing 60 percent or more of AN by weight. AN should also be handled in accordance with safe practices found in *NFPA 400 Hazardous Materials Code, Chapter 11.*

Building Design

- Store only in one-story buildings and buildings with no basements, unless the basement is open on one side.

- Use fire resistant walls within 50 feet of combustible building or materials.
- Flooring in storage and handling areas should be constructed of noncombustible material or protected from impregnation by AN.
- Avoid installing, or remove or close off any open drains, traps, tunnels, pits or pockets into which molten AN can flow and be confined in the event of fire.
- Buildings should be kept dry and free of water seepage through roofs, walls and floors.
- Have adequate ventilation or be constructed to self-ventilate in the event of a fire to avoid pressurization.
- Do not place AN into storage when the temperature of the product exceeds 130°F (54.4°C).

Storage in Bags, Drums or Other Containers

- Piles of bags, drums and other containers should be no closer than 36 inches below the roof or supporting beams.
- Bags should be stored no less than 30 inches from walls or partitions.
- Piles of bags, drums, and other containers should not exceed a height of 20 feet, width of 20 feet, and length of 50 feet, unless the building is of noncombustible construction or protected by automatic sprinklers.
- Maintain aisles of at least 3 feet width between piles.

Storage in Bulk

- Bins for storing bulk AN should be kept clean and free of materials, which could contaminate the material. Bins should not be constructed of galvanized iron, copper, lead or zinc unless suitably protected. Aluminum or wooden bins should be protected against impregnation by AN.
- Piles or bins must be adequately sized, arranged and moved periodically to minimize caking. Height or depth of piles shall be

limited by pressure-setting tendency of the product, but in no case should pile be higher than 36 inches below roof or supporting beams.

- Do NOT use dynamite, explosives or blasting agents to break up or loosen caked AN.
- Protect piles of AN from absorbing moisture from humid air by covering them with water-impermeable sheeting or using air conditioning.
- Do not store AN with organic chemicals, acids, or other corrosive materials, materials that may require blasting during processing or handling, compressed flammable gases, flammable and combustible materials or other contaminating substances. AN stores should be separated from incompatible substances by using separate buildings or 1 – hour fire resistant walls, or a minimum separation distance of 30 feet.

Fire Protection

- AN storage areas should be equipped with an automatic sprinkler system, or have an automatic fire detection and alarm system if the areas are not continuously occupied. This is especially important when the facility in question is close to the public surrounding the facility.
- Facilities should NOT store more than 2500 tons of bagged AN without an automatic sprinkler system.
- An automatic sprinkler system, if installed, should be provided in accordance with NFPA 13, Standard for the Installation of Sprinkler Systems.
- Suitable fire control devices such as hoses and appropriate portable fire extinguishers (AN is an oxidizer and not all fire extinguishers are appropriate) shall be provided throughout the warehouse and loading areas. Water supplies and fire hydrants should be available.
- Store AN fertilizer in separate buildings or separated by approved fire walls from organic, combustible or reactive materials, such as grains, wood or other organic materials, urea and urea compounds, flammable liquids or gases, corrosive acids, chlorates, chromates nitrites, permanganates or finely divided metals or sulfur.

- AN fertilizer should NOT be stored in the same building with explosives or blasting agents unless conditions in ATF's Table of Separation Distances of Ammonium Nitrate and Blasting Agents from Explosives and Blasting Agents, 27 CFR 555.220, are met.
- Prohibit smoking in AN storage areas.

We recommend that AN be stored in purpose-built facilities/buildings of non-combustible construction. Dust-producing organic materials, such as grain, seeds and sugar, should not be stored near AN. Some metal powders such as aluminum powder are equally dangerous. AN should be stored so as to ensure it is not contaminated by gasoline, diesel or other fuels, and is not subject to high heat (even in one small area of a large stockpile) or water infiltration.

COMMUNITY EMERGENCY PLANNING

What Should Communities Do to Understand and Develop a Plan for the Risk Associated with AN?

AN is a hazardous chemical covered under the OSHA Hazard Communication Standard. Therefore, facilities that handle and store AN are required by law to submit information regarding chemical hazards (including AN) to their State or Tribal Emergency Response Commission (SERC or TERC), Local Emergency Planning Committee (LEPC), and local fire department. This information must include the following:

1) a Safety Data Sheets (SDS) providing the chemical's hazard information and emergency response guidelines and
2) a Hazardous Chemical Inventory form that provides the quantity, storage types and locations of the AN at their facility.

We recommend that fire services visit any facility reporting AN, and that the conditions of storage and manner of handling be reviewed by fire service personnel. Fire service and other emergency responders should take note of the specific location(s), amounts and packaging of stored AN. Conditions of storage should be reviewed with the facility operator in light of the information provided in this document.

The LEPC in conjunction with the fire department should use this information to develop an emergency plan, in case of a fire or explosion involving AN or any other hazardous substance. The facility should consult with the LEPC to provide them the necessary information to develop the emergency plan, the elements of which should include:

- Identification of facilities and transportation routes of hazardous substances
- Description of emergency response procedures, on and off site
- Designation of a community coordinator and facility emergency coordinator(s) to implement the plan
- Outline of emergency notification procedures
- Description of how to determine the probable area and population affected by releases
- Description of local emergency equipment and facilities and the persons responsible for them
- Outline of evacuation plans
- A training program for emergency responders (including schedules)
- Methods and schedules for exercising emergency response plans

LEPCs should also ensure that members of the community (which would include potentially affected populations) are aware of the emergency plan and the actions they need to take if an accident occurs.

Local fire departments should use the information to determine what precautions they may need to take in responding to an accident at the facility and ensure the first responders have the appropriate training to respond to incidents involving AN.

Owners and operators of facilities holding AN are required to report the AN hazard to local response officials under the Emergency Planning and Community Right-to-Know Act (EPCRA). Unfortunately, that obligation is not universally understood, and so some facilities may fail to report. Fertilizer-grade AN is typically found at those businesses that provide direct logistical support to agriculture. This may include crop service operations, farm co-ops, grange stores and similar operations.

In the interest of community safety, it is often necessary and appropriate for first response officials to reach out to facility owners and operators, and determine if unreported risks are present in their community. Helping a neighbor, facility operator, or employer to understand and meet his obligations to the community and to workers is in everyone's best interest

EMERGENCY RESPONSE

Owner/operators of storage facilities should develop a site emergency response plan which includes:

- Coordination with local first responders
- Joint training with first responders if possible
- Employee training
- Community outreach
- Analysis of what may be at risk in a serious accident and appropriate planning
- Signs that clearly mark high hazard areas, safe areas, emergency contact numbers, firefighting equipment, and other essential area during an emergency response
- A site and area evacuation plan

Owners and operators of facilities holding AN have an obligation to ensure their community's first responders are aware of the hazards associated with the AN. Reliance on a report may not always be sufficient. Owners and operators should take a pro-active approach to reaching out to the emergency response officials in their location and ensuring that the hazards of AN are understood by the responders.

What Do Firefighters Need to Know When Responding to an Accident or Fire Involving AN?

Before responding to a fire involving AN, firefighters should ensure the community emergency response plan includes:

- AN hazard information and emergency response guidelines
- Quantity, storage types, and locations of AN at facilities in their community
- Specific response procedures; including a decision process to determine under which conditions a fire should be fought or whether the fire should be allowed to burn
- Evacuation procedures for the community
- Training requirements for all response personnel
- A schedule for exercising the response plan related to AN accidents

When responding to a fire where AN is stored; firefighters should:

- First consider if they can safety fight the fire or whether they should just let it burn, move to a safe location, and focus on evacuating nearby residents and preventing further safety issues for the surrounding community.

To determine whether or not it makes sense to fight the fire or to let it burn, firefighters and emergency responders should consider the following information:

- Firefighters should not fight an AN fire and everyone, including fire fighters, should be evacuated to a safe distance if they observe any of the following:
    - A fire involving AN is judged to be out of control;
    - The fire is engulfing the AN; or
    - Brown/orange smoke is detected, indicating the presence of nitrogen dioxide (which is toxic); or
    - A rapid increase in the amount/intensity of smoke or fire in the area of AN storage.
- If firefighters consider it safe and appropriate to respond to a fire involving AN, then the following information should be considered:
    - AN fires should be fought from protected locations or maximum possible distance. Approach a fire involving or close to AN from upwind to avoid hazardous vapors and toxic decomposition products. Self-contained breathing apparatus (SCBA) of types approved by the National Institute for Occupational Safety and Health (NIOSH) should be used to protect personnel against gases.

- o Use flooding quantities of water from a distance as promptly as possible. It is important that the mass of AN be kept cool and the burning be quickly extinguished. Keep adjacent fertilizers cool by spraying with large amounts of water. When possible and appropriate, only use unmanned hose holders or monitor nozzles.
- o Do NOT use steam, CO2, dry powder or foam extinguishers, sand or other smothering agents.7
- o Ensure maximum ventilation of the AN storage container as quickly as practical to prevent heat and pressure buildup. This is different than ensuring maximum ventilation of the entire building or structure where the AN is stored. Ventilation of the structure should be conducted only in a manner to limit fire spread and growth and should be minimized until a suppression water supply is established.
- o If practicable and safe to do so, attempt to prevent AN from entering the drains where explosive confinement could occur. Remember AN may be washed into drains by fire water, but it can also melt and flow without impetus from water.
- o Prevent or minimize contamination of water bodies or streams to reduce the potential for environmental effects.

INFORMATION RESOURCES

Codes and Standards

NFPA codes and Compressed Gas Association (CGA) standards are developed through a consensus standards development process approved by the American National Standards Institute. This process brings together volunteers representing various viewpoints and interests to achieve consensus on safety issues. These codes and standards are not binding but may be adopted by reference into laws or regulations. Users of the codes and standards should consult applicable federal, state and local laws and regulations.

NFPA has developed a code for storage of AN, including mixtures containing 60 percent or more by weight of AN, and a code for explosives that would apply to blasting agents and explosives containing AN. These codes are listed below:

NFPA 400 — Hazardous Materials Code, Chapter 11 - Ammonium Nitrate Solids and Liquids. (2013). Also see Annex A.11 in this document and Annex E: Properties and Uses of Ammonium Nitrate and Fire-Fighting Procedures.

NFPA 495 — Explosive Materials Code (2013).

National Fire Protection Association
1 Batterymarch Park
PO Box 9101
Quincy, MA 02169-7471
Phone: 800-344-3555 (toll free)
Website: http://www.nfpa.org/freeaccess

Safe Practices for the Production of Nitrous Oxide From Ammonium Nitrate, CGA G-8.4 (January 2013). Compressed Gas Association, Inc., Chantilly, VA http://www.cganet.com/customer/publication detail. aspx?id=G-8.4

GENERAL REFERENCES

Storing and Handling Ammonium Nitrate, INDG230 (First published 8/96, Reprinted 11/04). Health and Safety Executive (HSE), United Kingdom http://www.hse.gov.uk/explosives/ammonium/

Safe Storage and Handling of Ammonium Nitrate (AN), Technical Note 60, (28/02/2006), SafeWork, South Australia. http://www.safework.sa.gov.au/uploadedfiles/SSANStorage.T60.pdf

Safe Practice: Safe Storage of Solid Ammonium Nitrate. (2013). Resources Safety, Division of Mines and Petroleum, Government of Western Australia (WA), East Perth, WA. http://www.dmp.wa.gov.au/documents/Code of Practice/DGS COP StorageSolidAmmoniumNitrate. pdf

Guidance for the Storage, Handling and Transportation of Solid Mineral Fertilizers. (2007). European Fertilizers Manufacturers Association, Brussels, Belgium, www.efma.org

Guidance for the Safe Handling and use of Non-conforming Fertilizers and Related Materials (Producers). (2003). European Fertilizers Manufacturers Association, Brussels, Belgium, www.efma.org

Guidance for the Safe Handling and Use of Non-conforming Fertilizers and Related Materials for Fertilizer Importers, Distributors and Merchants. (2004). European Fertilizers Manufacturers Association, Brussels, Belgium, www.efma.org

Guidance for the Storage of Hot Ammonium Nitrate Solution. (2005). European Fertilizers Manufacturers Association, Brussels, Belgium, www.efma.org

Guidance for Compatibility of Fertilizer Blending Materials. (2006). European Fertilizers Manufacturers Association, Brussels, Belgium, www.efma.org

The above five guidance documents from European Fertilizers Manufacturers Association can be found on the following webpage: http://www.productstewardship.eu/site/index.php?id=259

Ammonium Nitrate and Mixed Fertilizers Containing Ammonium Nitrate, FM Global Property Loss Prevention Data Sheet 7-89. (April 2013). FM Global, Johnston, Rhode Island. http://www.fmglobal.com/page.aspx?id=04010200 Free access with registration

Ammonium Nitrate Handling, (2013). Bunn Fertiliser, Ltd. http://www.bunnfertiliser.com/infocentre/bunnhealthsafety/ammoniumnitratehandling/

Ammonium Nitrate, Industrial Grade, Technical Information. (2011) Dyno Nobel Inc. http://www.dynonobel.com/files/2010/04/1Ammonium Nitrate LomoDonora-Industrial.pdf

Ammonium Nitrate, Nutrient Source Specific (NSS) Fact Sheet, No. 22 International Plant Nutrition Institute, Norcross, GA http://www.ipni.net/publication/nss.nsf/0/67265A0AC9302CC5852579AF0076927A/$FILE/NSS22%20Amm%20Nit.pdf

Fire Protection Guide to Hazardous Materials, 14th edition. (2010). National Fire Protection Association, Quincy, MA.

Guide No. 140 for Oxidizers , *Emergency Response Guidebook.* 2012. US Dept. of Transportation, Pipeline and Hazardous Materials Safety Administration. http://www.phmsa.dot.gov/staticfiles/PHMSA/DownloadableFiles/Files/Hazmat/ERG2012.pdf

EPA Chemical Accident Investigation Report, Terra Industries, Inc., Nitrogen Fertilizer Facility, Port Neal, Iowa. (January, 1996). U.S. Environmental Protection Agency, Region 7, Emergency Response and Removal Branch, Kansas City, KS. http://www.epa.gov/emergencies/docs/chem/cterra.pdf

West Fertilizer Explosion and Fire. (2013). U.S. Chemical Safety Board http://www.csb.gov/west-fertilizer-explosion-and-fire-/

The National Safety Council has a data sheet *Ammonium Nitrate Fertilizer, Data Sheet I-699.* (1991) that discusses the health hazards, properties, and precautions for safe storage and handling of AN fertilizer.

National Safety Council
1121 Spring Lake Drive

Itasca, IL 60143-3201
Phone: (800) 621-7269 (toll free) or (630)-775-2199 (Library)
Website: http://www.nsc.org

The Fertilizer Institute (TFI) possesses information on various fertilizer products, including AN, and their uses.

The Fertilizer Institute
425 Third Street, SW, Suite 950
Washington, DC 20024 Phone: (202) 962-0490
Website: http://www.tfi.org

ResponsibleAg (RA) is a Fertilizer Code of Practice management system that helps facilities establish basic Environmental, Health, Safety and Security (EHS&S) performance practices. ResponsibleAg is a joint venture of the Agricultural Retailers Association (ARA) and The Fertilizer Institute (TFI). ARA also has a *First Responder Guidance* for use by agricultural retailers, LEPCs and local first responders. For more information, contact:

Agricultural Retailers Association
1156 15th Street, NW
Suite 500
Washington, D.C. 20005
Phone: 202-457-0825
Website: www.aradc.org

For more detailed information on the safe handling practices for storage of explosive materials which may contain AN, please consult the following Safety Library Publications (SLPs) developed by the Institute of Explosive Makers (IME).

Construction Guide for Storage Magazines, IME SLP No. 1 (September 2006).

The American Table of Distances, IME SLP No. 2 . (October 2011).

Suggested Code of Regulations for the Manufacture, Transportation, Storage, Sale, Possession, and Use of Explosive Materials, IME SLP No. 3. (October 2009).

Handbook for the Transportation and Distribution of Explosive Material, IME SLP No. 14. (April 2007).

Safety in the Transportation, Storage and Use of Explosive Materials, IME SLP No. 17 (October 2011).

Recommendations for the Transportation of Explosives, Division 1.5, Ammonium Nitrate Emulsions, Division 5.1, Combustible Liquids, Class 3, and Corrosives, and Liquids, Class 8 in Bulk Packaging, IME SLP No. 23. (October 2011).

Explosives Manufacturing and Processing Guide to Safety Training, IME SLP No. 25. (May 2011).

SLPs are available at http://www.ime.org/ecommerce/products.php?category id=13

Institute of Makers of Explosives (IME) 1120 Nineteenth St. N.W. Suite 310 Washington, DC 20036-3605

Phone: (202) 429-9280

Website: www.ime.org

SAFEX International is an industry group whose members manufacture civil or military explosives or technical grade ammonium nitrate (TGAN). TGAN is generally in the form of porous prills and is used in the manufacture of commercial explosives. SAFEX has published a guide for safe storage of TGAN listed below that is available to its members. https://www.safex-international.org/ index.php

Good Practice Guide: Storage of Solid Technical Grade Ammonium Nitrate. (March 2011). International Working Group on Ammonium Nitrate, SAFEX International. SAFEX Good Explosive Practice Series, GPG 02 rev. 1

STATUTES AND REGULATIONS

Statutes and regulations applicable to the manufacture of or processes involving AN, are listed below.

Clean Air Act Accident Prevention- General Duty (EPA)

Section 112(r) of the Clean Air Act (CAA) focuses on prevention of chemical accidents. Under this provision of the CAA, all facilities with regulated substances or other extremely hazardous substances have a general duty to prevent and mitigate accidental releases. Under Section 112(r)(1), the general duty is :

- to identify hazards ...using appropriate hazard assessment techniques, to design and maintain a
- safe facility taking such steps as are necessary to prevent releases, and to minimize the consequences of accidental releases which do occur.

This general duty applies to facilities producing, processing, handling or storing extremely hazardous substances. While not a regulated substance, AN may be considered extremely hazardous under certain circumstances.

Clean Air Act- Risk Management Program (EPA) and Process Safety Management (OSHA)

In 1990, amendments to the CAA authorized the EPA's Risk Management Program (RMP) Rule (40 CFR Part 68) under section 112(r), and required the Occupational Health and Safety Administration (OSHA) to issue the Process Safety Management Program (PSM) rule. Both rules serve to prevent chemical accidents. The RMP focuses on prevention and mitigation of accidental releases of listed toxic and flammable substances. Requirements under the RMP rule include development of a hazard assessment, a prevention program, and an emergency response program. While AN is not a listed substance subject to the RMP, chemicals used in the production of AN are included on the RMP list, making the process producing AN potentially subject to the RMP. Certain processes using AN may also involve RMP listed substances. For more information about RMP regulations, see http://www.epa.gov/emergencies/content/rmp/index.htm

OSHA's Process Safety Management (PSM) Standard establishes requirements intended to protect employees by preventing or minimizing the consequences of chemical accidents involving highly hazardous chemicals (29 CFR 1910.119). Although AN is not covered by the PSM standard, the production or use of AN may involve PSM listed chemicals in excess of thresholds. Manufacture of explosives, which may involve AN, is also covered by the PSM standard. For more information about OSHA's PSM standard see https://www.osha.gov/SLTC/processsafetymanagement/index.html

Emergency Planning and Community Right-to-Know Act (EPA)

The Emergency Planning and Community Right-to-Know Act (EPCRA), requires information on the presence of hazardous chemicals above designated threshold quantities at regulated facilities be provided to state and local emergency planning authorities. This information facilitates development of emergency response plans required by section 303 of EPCRA, enhances community awareness of chemical hazards and help first responders to respond to chemical accidents. The chemicals covered under these requirements are a specific list of chemicals known as Extremely Hazardous Substances (EHSs) found at 40 CFR Part 355 Appendices A and B and any chemicals that meet the criteria as hazardous chemicals under OSHA's Hazard Communication Standard. AN is not an EHS but is considered a hazardous chemical (oxidizer) and therefore subject to the EPCRA provisions described below.

Section 311 of EPCRA requires facilities to submit Safety Data Sheets for the EHSs and hazardous chemicals to their State or Tribal Emergency Response Commission (SERC or TERC), Local Emergency Planning Committee (LEPC) and local fire department. Section 312 requires facilities to submit annually to their SERC or TERC, LEPC, and local fire department, Hazardous Chemical Inventory forms for these chemicals. The SDS provides the chemical's hazard information and emergency response guidelines and the Hazardous Chemical Inventory form provides the quantity, storage types and locations of the chemical at their facility. Regulations covering these requirements are found at 40 CFR Part 370.

Section 311(e)(5) of EPCRA does not include the following as a hazardous chemical: any substance used in routine agricultural operations or a fertilizer held for sale by a retailer to the ultimate customer. At fertilizer distributors, AN is commonly blended with other chemicals to produce a fertilizer mix according to customer specifications. Any AN that is mixed or formulated with other chemicals by facilities is not covered by the Section 311(e)(5) exemption. The exemption was intended to apply only to retailers of the substance, not to manufacturers and wholesalers – who typically have large quantities of fertilizers, and may use and manufacture a wide range of chemical compounds. These manufacturers and wholesalers can present significant risks that need to be addressed by emergency response authorities. For more information about EPCRA hazardous chemical reporting, see http://www.epa.gov/emergencies/content/epcra/index.htm

Environmental Protection Agency (EPA) Phone: (800) 424-9346 or (703) 412-9810 Website: http://www.epa.gov

Explosives and Blasting Agents Standards (OSHA)

In addition to the PSM program described above, the Occupational Safety and Health Administration (OSHA) regulates the manufacture, keeping, having, storage, sale, transportation, and use of explosives and blasting agents under its Occupational Safety and Health Standards for explosives and blasting agents (29 CFR 1910.109). Blasting agents are frequently formulated with AN. For more information about OSHA's standards covering explosives and blasting agents, including ammonium nitrate and storage of all grades of ammonium nitrate, see https://www.osha.gov/pls/oshaweb/owadisp.show document?p id=9755&p table=STANDARDS

Hazard Communication Standard (OSHA)

OSHA's Hazard Communication Standard (HCS) at 29 CFR 1910.1200 requires chemical manufacturers and importers to evaluate the hazards of the chemicals they produce or import, and prepare labels and Safety Data Sheets (SDS) to convey the hazard information to their downstream customers. All employers with hazardous chemicals in their workplaces must have labels and safety data sheets for their exposed workers, and train them to handle the chemicals appropriately. AN is a hazardous chemical covered under the HCS. The HCS is now aligned with the Globally Harmonized System of Classification and Labeling of Chemicals (GHS). Employers are required to train workers by December 1, 2013 on the new labels elements and safety data sheets format to facilitate recognition and understanding. For more information, see http://www.osha.gov/dsg/hazcom/index.html

Occupational Safety and Health Administration Phone: (800) 321- OSHA (6742) Website: http://www.osha.gov

Chemical Facility Anti-Terrorism Standards (DHS)

The Department of Homeland Security (DHS)'s Chemical Facility Anti-Terrorism Standards (CFATS) program applies to facilities that possess threshold quantities of certain types of ammonium nitrate. Facilities in possession of Chemicals of Interest (listed in 6 CFR Part 27 Appendix A) exceeding specific threshold quantities are required to complete a "Top-Screen" questionnaire to identify the types and quantities of Chemicals of Interest the facility possesses. For ammonium nitrate at any concentration (with more than 0.2% combustible substances, including any organic substance calculated as carbon, to the exclusion of any other added substance) the Screening Threshold Quantity for risk of release is 5,000 pounds and for risk of theft is 400 pounds. This same form of ammonium nitrate is also classified by DOT as a Division 1.1 explosive. For solid ammonium nitrate, with a minimum concentration of 33% or greater and a nitrogen concentration of 23% nitrogen or greater, the Screening Threshold Quantity for risk of theft is 2,000 pounds. The CFATS program, first established under Section 550 of the 2007 DHS Appropriations Act, identifies and regulates high-risk chemical facilities to ensure they have security measures in place to reduce the risks associated with these chemicals. CFATS regulations are found in 6 CFR Part 27.

Based on the Top-Screen, if DHS initially determines the facility to be high-risk, the facility must complete and submit a Security Vulnerability Assessment, which is then reviewed by DHS to make a final determination on whether the facility is high-risk. Facilities receiving a final high-risk determination must develop and submit for DHS's review, a Site Security Plan (SSP), or alternatively, an Alternative Security Program, that describes the specific security measures the facility will utilize to meet the 18 applicable risk-based performance standards under CFATS. The agency must then conduct an inspection to help determine whether or not the facility's SSP should be approved. For more information about CFATS program, see http://www.dhs.gov/chemical-facility-anti-terrorism-standards

Hazardous Materials (DOT)

The Department of Transportation (DOT) regulates transportation of AN under its Hazardous Materials Regulations.

The following forms of ammonium nitrate are listing in the DOT Hazardous Materials Table (49 CFR 172.101) with their Hazard Class or Division:

Ammonium nitrate based fertilizer, 5.1

Ammonium nitrate based fertilizer, 9 (when transported by vessel or aircraft)

Ammonium nitrate emulsion or Ammonium nitrate suspension or Ammonium nitrate gel, intermediate for blasting explosives, 5.1

Ammonium nitrate-fuel oil mixture containing only prilled ammonium nitrate and fuel oil, 1.5D Ammonium nitrate, liquid (hot concentrated solution), 5.1

Ammonium nitrate, with more than 0.2 percent combustible substances, including any organic substance calculated as carbon, to the exclusion of any other added substance, 1.1D

Ammonium nitrate, with not more than 0.2% total combustible material, including any organic substance, calculated as carbon to the exclusion of any other added substance, 5.1

Explanation of Hazard Class Numbers

1.1 - Explosives (with a mass explosion hazard) A mass explosion is one which affects almost the entire load instantaneously.

1.5 - Very insensitive explosives; blasting agents

5.1 - Oxidizer

9 - Miscellaneous Hazard Material

DOT also requires security plans for persons offering for transportation or transporting any quantity of a Division 1.1 or 1.5 material containing ammonium nitrate or large bulk quantities (greater than 6,614 lbs or 792 gals) of ammonium nitrate, ammonium nitrate fertilizers, or ammonium nitrate emulsions, suspensions, or gels. The security plan must conform to requirements in 49 CFR 172.800.

Department of Transportation

Phone: (202) 366-5580 - Public Information Website: http://www.dot.gov

Explosives Regulations (ATF)

The Bureau of Alcohol, Tobacco, Firearms, and Explosives (ATF) of the Department of the Justice regulates the importation, manufacture, distribution, and storage of explosive materials including blasting agents and other explosive materials containing AN. ATF's explosives regulations, 27 CFR Part 555, can be located at http://www.atf.gov/regulations-rulings/regulations/index.html

Bureau of Alcohol, Tobacco, Firearms, and Explosives Phone: (202) 648-7120 Website: http://atf.gov

For More Information, Contact:

The Superfund, TRI, EPCRA, Risk Management Program, and Oil Information Center
(800) 424-9346 or (703) 412-9810
TDD (800) 553-7672 or (703) 412-3323

End Notes

[1] The statements in this document are intended solely as guidance. This document is not a substitute for EPA, OSHA, ATF or other agencies' regulations, nor is it a regulation itself. It cannot and does not impose legally binding requirements on the agencies, states, or the regulated community. The measures described in this document may not apply to a particular situation based upon the circumstances. This guidance does not represent final agency action and may change in the future, as appropriate.

[2] A blasting agent is any material or mixture, consisting of a fuel and oxidizer, intended for blasting, not otherwise classified as an explosive and in which none of the ingredients are classified as an explosive, provided that the finished product, as mixed and packaged for use or shipment, cannot be detonated by means of a No. 8 test blasting cap when unconfined (see 29 CFR 1910.109(a)(1))

[3] An oxidizer is a material that readily yields oxygen or other oxidizing gas or that reacts readily to promote or initiate combustion of combustible materials.

[4] The precise quantity and form of AN has not been definitively established. We intend to update this advisory as we learn more about the incident and as we identify additional best practices.

[5] Explosive means any substance or article, including a device, which is designed to function by explosion (i.e., an extremely rapid release of gas and heat) or which, by chemical reaction within itself, is able to function in a similar manner even if not designed to function by explosion (see 49 CFR 173.50(a)).

[6] AN-based materials that are DOT Hazard Class 1 sensitive (explosives or blasting agents) must be handled and stored in accordance with requirements of OSHA's Standard for Explosives and Blasting Agents (29 CFR 1910.109) and ATF's Table of Separation Distances of Ammonium Nitrate and Blasting Agents from Explosives or Blasting Agents (27 CFR 555.220) Facilities should also follow the NFPA 495- Explosive Materials Code, where applicable.

[7] Keep in mind that ammonium nitrate is an oxidizer – that is – it provides its own oxygen and once combustion begins, it cannot be smothered. Moreover, the combination of heat and confinement will accelerate combustion, perhaps to the point of detonation.

In: Ammonium Nitrate Facilities
Editor: Allegra Ward

ISBN: 978-1-63321-716-4

Chapter 4

Testimony of Rafael Moure-Eraso, Chairman, U.S. Chemical Safety Board. Hearing on "Oversight of Federal Risk Management and Emergency Planning Programs to Prevent and Address Chemical Threats, Including the Events Leading Up to the Explosions in West, TX and Geismar, LA"*

Chairman Boxer, Ranking Member Vitter, and distinguished members of the Committee – thank you for the opportunity to testify before you this morning. I am Dr. Rafael Moure-Eraso, and I am providing this testimony in my capacity as chairperson of the U.S. Chemical Safety Board, or CSB.

The CSB is an independent federal agency that investigates major chemical accidents and hazards, and develops safety recommendations to prevent their recurrence in the future. The Board is a non-regulatory, scientific, investigative agency. It has an annual budget, after the sequester, of $10.6 million and approximately 42 employees. In addition to investigations, safety studies, and recommendations, we do extensive outreach to companies

* This is an edited, reformatted and augmented version of testimony presented June 27, 2013 before the Senate Committee on Environment and Public Works.

and other organizations to inform them of our findings. Companies throughout the U.S. and the world use the information and recommendations developed by the CSB to help create what we hope are safer workplaces.

Congress frequently calls upon the CSB to investigate the root causes of some of the most complex and tragic industrial accidents across the country. Currently the CSB is involved in investigations of the Deepwater Horizon blowout in the Gulf of Mexico, the 2010 Tesoro refinery fire in Washington State, the 2012 Chevron refinery fire in California, and many other cases. Over the past two months, the CSB has begun investigations of the devastating explosion at West Fertilizer in West, Texas, on April 17, and the June 13 explosion at Williams Olefins in Geismar, Louisiana.

I will summarize the status of these two investigations and our preliminary findings, and then present some general thoughts on how the oversight of chemical safety might be improved.

West Fertilizer

West Fertilizer was a small retail distribution center that served farmers in the surrounding community and had approximately 15 employees. The facility was built in 1961, and at the time of the incident had a handful of buildings, including a warehouse where fertilizers and other materials were stored. The current owner, who operated an adjacent seed business, purchased the facility from liquidation in 2004.

No manufacturing occurred at the site, only blending of fertilizers for retail customers.

Fertilizers such as ammonium nitrate and anhydrous ammonia were delivered to the site by rail car or truck. The ammonium nitrate, a granular solid, was stored in the facility's fertilizer warehouse building in wood-framed bins with wooden walls. Both the warehouse building and the bins were constructed of combustible wooden material, and the building also contained significant quantities of combustible materials such as seeds stored near the bins of ammonium nitrate. The building had no automatic sprinkler or fire suppression features.

The facility straddles the city limit in the northeast section of West, Texas. When it was first built, the area was rural and there were few other structures nearby. Over time, many residences, a nursing home, an apartment complex, a high school, and an intermediate school were constructed within a 2000-foot radius of West Fertilizer.

On the evening of April 17, a fire of undetermined origin broke out at the facility, which had already closed for the day and was unattended. At 7:30 p.m. the fire was observed and reported to 9-1-1 dispatchers, who deployed the community's volunteer firefighting force with four pieces of equipment. Firefighters found the warehouse building in flames and were in the process of extending hoses to fight the fire, and were applying some water to the blaze. Although the firefighters were aware of the hazard from the tanks of anhydrous ammonia as a result of previous releases, they were not informed of the explosion hazard from the approximately 60 tons of fertilizer grade ammonium nitrate inside the warehouse.

At about 7:50 p.m., while firefighters were positioned nearby, the ammonium nitrate suddenly detonated. A shock wave, traveling faster than the speed of sound, crushed buildings, flattened walls, and shattered windows. Innumerable projectiles of steel, wood, and concrete – some weighing hundreds of pounds – were hurled into neighborhoods. Twelve firefighters and emergency responders were killed. At least two members of the public died as well. More than 200 were injured. If this incident had occurred earlier in the day, many more people might have been killed or injured.

Residents of the West Rest Haven nursing home were severely affected, and according to nursing home officials 14 patients have passed away since the April 17 explosion, dying at twice the expected rate. The nursing home itself was destroyed, as was the apartment complex across the street. Two large schools – the high school and the intermediate school – were structurally damaged beyond repair and will be torn down, and a third school was also badly damaged. Because of the hour of day, all the schools were unoccupied. Had the explosion taken place during the day, severe casualties could have occurred in the intermediate school, which was devastated by both blast and fire. Post-explosion damage assessments indicate that it would have been difficult for children and others to escape from the building. The CSB is currently evaluating the vulnerability of this structure, to understand the potential consequences if the explosion had occurred when children were present and to inform future siting decisions.

Nearly 200 homes were severely damaged or destroyed, a sizeable fraction of all the houses in West. Financial damage is still being assessed, but the cost to rebuild the schools alone will reportedly approach $100 million. Some reports suggest total damages to the town may exceed $230 million, an unimaginable blow to a town of just 2800 residents – more than $80,000 for each man, woman, and child living in West.

CSB INVESTIGATION

A large CSB investigation team was assembled in West the day after the incident, on April 18. To date the CSB has conducted detailed interviews of about 30 witnesses, and has issued approximately 13 document requests to West Fertilizer, contract firms, hospitals, and regulators.

The CSB has also engaged external experts in blast reconstruction, fire codes and fire protection, and explosion mechanisms.

West Fertilizer and other companies have cooperated fully with the investigation. The CSB has also received outstanding cooperation from the mayor of West and its police and fire departments, and from other local agencies. The investigation has faced significant challenges as well, since the accident site was treated as a criminal scene for approximately five weeks after April 17 and was extensively altered during that time period, including the removal of most surviving physical evidence.[1]

I visited West, Texas, on May 2, just a couple of weeks after the explosion. The damage to homes, schools, and businesses was almost beyond imagination – even by the standards of large-scale chemical disasters. My heart goes out to the people of West, as they work to rebuild their proud and historic community. But I can assure you that it will be years before even the physical scars of this terrible explosion begin to fade.

Ammonium nitrate (AN) is a crop nutrient that represents about 2% of the total applied nitrogen fertilizer in the U.S. It is used primarily on pasture and citrus; its use has been declining in recent years as security concerns have increased since the Oklahoma City bombing in 1995. Ammonium nitrate is a strong oxidizer that reacts energetically with organic materials; it is also reactive by itself and capable of a runaway decomposition reaction and detonation under certain conditions.

Ammonium nitrate has historically been involved in some of the most severe chemical accidents of the past century, including disastrous explosions in the United States, Germany, and France. Two of these accidents – in Oppau, Germany, in 1921 and in Texas City, Texas, in 1947 – each killed 500 or more people. Additional safeguards were adopted following the Texas City disaster, such as avoiding contamination with petroleum-based materials that sensitize AN. These changes are credited with reducing the risk of a mass explosion of AN, but the risk of detonation was not eliminated. In September 2001, for example, a large AN explosion occurred at a factory in Toulouse, France, killing 30, injuring thousands of others, and damaging up to 30,000 buildings. Other serious AN-related accidents have occurred in the U.S. and other countries over the years.

Heat, fire, shock, confinement, and contamination are all factors that can sensitize ammonium nitrate to detonation. To quote from a comprehensive 1985 review of the hazards of AN:

> The main thrust of the safety precautions recommended in most literature is the minimization of the most likely hazard, namely, the risk of fire. Ammonium nitrate should not be stored where it can be affected by any source of heat or by combustible materials.[2]

As simple as this sounds, this principle has not been fully adopted across the U.S., and was not implemented at West Fertilizer.

The CSB has made the following observations and preliminary findings to date, which are subject to further revision and development as the investigation unfolds:

1. The explosion at West Fertilizer resulted from an intense fire in a wooden warehouse building that led to the detonation of approximately 30 tons of AN stored inside in wooden bins. Not only were the warehouse and bins combustible, but the building also contained significant amounts of combustible seeds, which likely contributed to the intensity of the fire. According to available seismic data, the explosion was a very powerful event.
2. Whether additional factors such as material characteristics, shock, or contamination contributed to the incident remains to be determined. Company employees described a PVC plastic pipe that was located directly above the AN bin that detonated, and likely would have been melted by the fire. Additionally, large amounts of potentially flammable anhydrous ammonia were stored along the southern edge of the warehouse building.
3. The building lacked a sprinkler system or other systems to automatically detect or suppress fire, especially when the building was unoccupied after hours. By the time firefighters were able to reach the site, the fire was intense and out of control. Just 20 minutes after the first notification to the West Volunteer Fire Department, the detonation occurred.
4. Both National Fire Protection Association (NFPA) and the International Code Council (ICC), private organizations that develop

fire codes that are widely applied across the U.S., have written code provisions for the safety of ammonium nitrate. Many of these safety provisions are quite old [3] and appear to be confusing or contradictory, even to code experts, and are in need of a comprehensive review in light of the West disaster and other recent accidents. For example the ICC's International Fire Code directs users to a defunct code for ammonium nitrate (NFPA 490, last issued in 2002) rather than the current code, known as NFPA 400.

5. The existing fire codes do contain some useful provisions; for example the codes do require a fire resistant barrier between AN and any stored flammable or combustible materials and have provisions to avoid AN confinement and promote ventilation during fire conditions. However, even the most current NFPA 400 standard allows AN to be stored in wooden buildings and in wooden bins, and does not mandate automatic sprinkler systems unless more than 2500 tons of AN is being stored – vastly more than the approximately 30 tons that was sufficient to devastate much of the town of West. In addition, the standard contains a "grandfathering" provision that allows existing buildings that were constructed prior to code adoption – and fail to meet all of its provisions – to continue in use.
6. Texas has not adopted a statewide fire code, and state law actually prohibits most smaller rural counties from adopting a fire code. McLennan County, where the West facility was located, had not adopted a fire code, although it technically had the authority to do so because of its proximity to the more populous Bell County. The West Fertilizer facility was thus not required to follow any NFPA or ICC recommendations for the storage of AN.
7. Although some U.S. distributors have constructed fire-resistant concrete structures for storing AN, fertilizer industry officials have reported to the CSB that wooden buildings are still the norm for the distribution of AN fertilizer across the U.S.
8. Industry has developed other forms of ammonium nitrate that are reported to reduce or eliminate the risk of accidental detonation. For example, compounding the ammonium nitrate with calcium carbonate (limestone) "practically eliminates any risk of explosion its storage, transportation, and handling," while preserving the AN's nutritive value.[4] Calcium ammonium nitrate fertilizers have been widely used in Europe. Ammonium sulfate nitrate also has been found to be non-explosive provided the percentage of AN is held below about 37%.[5]

9. The federal OSHA standard for "Explosives and Blasting Agents" (29 CFR 1910.109) does have requirements for ammonium nitrate fertilizer; its provisions are similar to the NFPA codes. Unlike the NFPA codes – which West was not legally required to follow under any fire code – the OSHA standard would have applied. Like NFPA, however, the OSHA standard does not prohibit wooden bins or wooden construction, and does not require sprinklers unless more than 2500 tons of AN is present. However, OSHA public records indicate that OSHA last inspected the facility in 1985, and no citations were issued under the "Explosives and Blasting Agents" standard.
10. OSHA's Process Safety Management standard (29 CFR 1910.119) or PSM was adopted in 1992 and is designed to prevent catastrophic workplace incidents involving highly hazardous chemicals. PSM requires companies to have a variety of management elements to prevent catastrophic incidents, such as conducting hazard analyses and developing emergency plans. Ammonium nitrate is not, however, one of the listed chemicals that triggers PSM coverage. The PSM standard also contains an exemption for retail facilities.
11. The EPA's Risk Management Program rule (40 CFR Part 68) or RMP was adopted in 1996 and is designed to prevent catastrophic offsite and environmental damage from extremely hazardous substances. As the name suggests, the rule requires covered facilities to develop a Risk Management Plan, implement various safety programs, and analyze offsite consequences from potential accidents. Once again, however, ammonium nitrate is not one of the listed chemicals that triggers RMP coverage. West Fertilizer was RMP-covered due to its stored ammonia, and the company's offsite consequence analysis considered only the possibility of an ammonia leak, not an explosion of ammonium nitrate.
12. OSHA considered adding ammonium nitrate along with other highly reactive chemicals to its list of PSM-covered substances in the late 1990's. However, this proposal was shelved in 2001. In developing the RMP regulation, the EPA did not explicitly include explosives or reactive chemicals in the list of covered chemicals. In 2002, the CSB issued a study on reactive hazards, identifying 167 prior reactive incidents (including a 1994 explosion at an ammonium nitrate manufacturer). The Board recommended that both OSHA and EPA expand their standards to include reactive chemicals and hazards. However, neither agency has yet acted upon the recommendations.

13. No federal, state, or local standards have been identified that restrict the siting of ammonium nitrate storage facilities in the vicinity of homes, schools, businesses, and health care facilities. In West, Texas, there were hundreds of such buildings within a mile radius, which were exposed to serious or life-threatening hazards when the explosion occurred on April 17.
14. West volunteer firefighters were not made aware of the explosion hazard from the AN stored at West Fertilizer, and were caught in harm's way when the blast occurred. NFPA recommends that firefighters evacuate from AN fires of "massive and uncontrollable proportions." Federal DOT guidance contained the Emergency Response Guidebook, which is widely used by firefighters, suggests fighting even large ammonium nitrate fertilizer fires by "flood[ing] the area with water from a distance." However, the response guidance appears to be vague since terms such as "massive," "uncontrollable," "large," and "distance" are not clearly defined. All of these provisions should be reviewed and harmonized in light of the West disaster to ensure that firefighters are adequately protected and are not put into danger protecting property alone.
15. While U.S. standards for ammonium nitrate have apparently remained static for decades, other countries have more rigorous standards covering both storage and siting of nearby buildings. For example, the U.K.'s Health and Safety Executive states in guidance dating to 1996 that "ammonium nitrate should normally be stored in single storey, dedicated, well-ventilated buildings that are constructed from materials that will not burn, such as concrete, bricks or steel."6 The U.K. guidance calls for storage bays "constructed of a material that does not burn, preferably concrete."
16. CF Industries, a principal manufacturer of AN that was one of the suppliers to West, also recommends more rigorous safeguards in its Material Safety Data Sheet (MSDS) for the chemical. In the section entitled "Handling and Storage," CF recommends that "Storage construction should be of non-combustible materials and preferably equipped with an automatic sprinkler system."[7] Although companies are required to issue MSDS's, the recipients of this information like West Fertilizer are not obligated to follow the recommended safety precautions. West lacked these safeguards.
17. The Bureau of Alcohol, Tobacco, Firearms, and Explosives (ATF) has regulations for ammonium nitrate used as an explosive but these do

not apply to ammonium nitrate used as fertilizer. The U.S. Department of Homeland Security has reporting requirements for companies that have a threshold amount of fertilizer grade ammonium nitrate. However, the authority of DHS is to require security measures to protect against theft, diversion, or other intentional acts; DHS does not regulate the safety of ammonium nitrate to prevent conditions leading to accidental detonation.

18. The Emergency Planning and Community Right-to-Know Act of 1986 (EPCRA) contains an exemption from hazardous chemical reporting for "fertilizer held for sale by a retailer to the ultimate customer." The EPA has interpreted this provision as not applying to firms, like West, that make custom blends of bulk fertilizer for customers' use. In 2012, West Fertilizer filed an EPCRA Tier II report with the McLennan County Local Emergency Planning Committee (LEPC). West reported the presence of up to 270 tons of ammonium nitrate, as well as anhydrous ammonia, at the site. The company did not provide the LEPC or the West Fire Department with an ammonium nitrate MSDS indicating the material's hazards, nor does EPCRA automatically require that information to be provided. There is no indication that West's filing with local authorities resulted in an effort to plan for an ammonium nitrate emergency.

It is important to bear in mind the limitations on Local Emergency Planning Committees that operate in communities around the country. While these committees are required to exist under EPCRA, they are largely staffed by either volunteers or local officials who likely have many collateral duties. The law did not establish any funding stream for the LEPC's, and they do not have any regulatory authority over chemical facilities. Their fundamental role is in emergency preparedness and coordination. The primary responsibility for developing and enforcing safety standards belongs to other federal and state agencies.

To summarize, the safety of ammonium nitrate fertilizer storage falls under a patchwork of U.S. regulatory standards and guidance – a patchwork that has many large holes. Specifically, the CSB has not identified any U.S. standards or guidance that prohibit or discourage many of the factors that likely contributed to the West disaster. Combustible wooden buildings and storage bins are permitted for storing AN across the U.S. – exposing AN to the threat of fire. Sprinklers are generally not required unless very large quantities of AN are being stored or fire authorities order sprinklers to be installed.

Federal, state, and local rules do not prohibit the siting of AN storage near homes and other vulnerable facilities such as schools and hospitals.

The CSB has had a number of discussions with fertilizer industry representatives since April 17, including officials from The Fertilizer Institute and the Agricultural Retailers Association. We believe the industry has a strong and sincere interest in learning from the tragedy in West and taking steps to prevent future incidents involving ammonium nitrate, including the development of new audit tools and product stewardship programs. I applaud these efforts and encourage these organizations to draw upon the best science as well as the strongest safety recommendations from the U.S. and overseas, to ensure that U.S. fertilizer firms are applying the highest safety standards available anywhere in the world.

These voluntary programs should complement a thorough effort by the federal government to review and improve the comprehensive safety oversight of ammonium nitrate fertilizer distribution. The time for that effort is now.

Williams Olefins Explosion

On June 13, an explosion and fire occurred at the Williams Olefins plant in Geismar, Louisiana. This plant produces ethylene and propylene, which are raw materials for common plastics, and employs over one hundred workers. At the time of the incident, hundreds of contract workers were also present at the site for a major expansion project.

The CSB deployed a team of seven to the site, and CSB investigators have had a continuous presence in Geismar since June 15. The team has interviewed at least 28 witnesses and has reviewed documents and other information obtained from the company. Williams Olefins and other companies at the site have provided excellent cooperation with the investigation.

The incident involved a large distillation tower that processes propylene, propane, and other highly flammable hydrocarbons. The equipment was in normal operation on June 13. At 8:36 a.m. there was a sudden catastrophic failure involving a heat exchanger and associated piping attached to the distillation tower. The steel shell of the heat exchanger ripped open, and piping detached where it connected to the tower. The exact sequence and cause of these events remains to be determined.

In any event, there was a large-scale release of propylene, propane, and other hydrocarbons from multiple release points, forming a vapor cloud more than 200 feet high that is visible in surveillance video from the site. Within four seconds the vapor cloud ignited. Two Williams employees were fatally burned and approximately 105 other Williams employees and contractors were injured. The resulting fire burned for over four hours.

All of us at the CSB offer our deepest condolences and prayers for the families of the victims and for the injured. We are committed to a thorough investigation to determine why this horrible accident occurred.

CSB investigators have surveyed the scene from ground level and from the air, but currently the immediate area of the ruptured equipment remains too hazardous for entry due to overhanging debris. During the course of this week the area will be made safe for human entry, and this will allow investigators to observe the positions of key valves and obtain other important information. In addition we plan to recover and perform metallurgical tests on the heat exchanger and other piping. This testing will help determine whether the equipment that failed had weakened or deteriorated prior to the rupture, or some other factors were at play.

We are also working with the company to recover electronic control system data that will reveal process conditions at the time of the incident, such as material flows, pressures, and temperatures as well as valve positions. These data will also be important to understanding what occurred.

The assessment of the site and equipment is occurring in close coordination with federal OSHA inspectors. Within a few days of the incident, the CSB, OSHA, and the company entered into a written site and evidence control agreement to ensure that the evidence at the site is properly preserved in as-found condition, and all parties participate in the identification and testing of evidence. So far it has been a good model for how all incident sites should be handled.

CSB Investigative Capacity

The recent tragedies in West and Geismar have further taxed the CSB's already overstretched staffing and resources. When the Congress requested that the CSB conduct a root-cause investigation of the Deepwater Horizon blowout, we informed Congress that this vital work would have unavoidable adverse effects on many other cases the CSB had already begun. The CSB already faced a record backlog of cases in 2010, when I became the chair. Not only have these adverse effects occurred, but Transocean – the operator of the Deepwater Horizon – has engaged in a lengthy legal challenge to the CSB's authority to investigate the incident. On April 1, 2013, a federal district court in Houston ruled completely in the CSB's favor and confirmed our offshore jurisdiction, but Transocean has indicated its intention to appeal the decision and seek a stay of enforcement. This unfortunate legal situation has continued to delay the CSB's access to many documents and witnesses relevant to the investigation of the blowout.

The West and Geismar investigations have very significant financial costs associated with them and West in particular has required the diversion of a very large percentage of CSB's investigators, who already had many months of work in the pipeline ahead of them when the tragedy struck. I would like to engage in a discussion with the Committee over the coming weeks about the impact of these new investigations on the CSB's capacity to finish existing investigations – many of which have important stakeholders who have already been waiting a long time for answers. I also wish to notify the Committee that I believe the CSB has no capacity at this point to undertake any new investigative work, beyond what has already been promised and begun.

Possible Approaches for Reducing Risk

Since the CSB was established in 1998, the Board has made a number of safety recommendations for improving the oversight of facilities that handle hazardous substances. The CSB has made a number of recommendations to the Environmental Protection Agency, including the above-mentioned recommendation to broaden the application of the Risk Management Program to encompass reactive hazards that could have an impact on communities. The CSB has also recently recommended that the EPA strengthen the safety provisions for disposing of hazardous waste; this followed a recent tragedy in Hawaii where five federal subcontractors were killed disposing of illegal fireworks seized by the government.

In another recent case, the CSB urged the EPA to make greater use of its general duty clause authorities under the Clean Air Act by warning operators of their responsibility to safeguard remote oil and gas production sites; the CSB investigation found that 44 members of the public – children and young adults – died in explosions at these unsecured hazardous sites.

The Board has made a number of safety recommendations to OSHA as well. Among the improvements we have sought are a new regulatory standard for combustible dust; broadening the PSM standard to cover reactive chemicals and atmospheric storage tanks and to require more effective management of change reviews; modernization of standards for acetylene and compressed gases; and developing a new safety standard for fuel gases.

The majority of the CSB's recommendations have not been directed to federal regulators but rather to other organizations around the country, including state and local governments, labor unions, trade associations, and the bodies like the ICC and NFPA that are responsible for developing consensus standards. The overall acceptance rate for CSB recommendations now exceeds 70%, and we track all recommendations to completion.

Improved enforcement efforts are just as important as having effective standards. In the CSB's 2007 report on the explosion at BP's Texas City refinery, the Board called for OSHA to expand its enforcement of process safety requirements by "hiring or developing a sufficient cadre of highly trained and experienced inspectors." The Board report observed that there were few comprehensive OSHA inspections of refineries and other chemical sites, and OSHA had only a handful of inspectors with industrial process experience. By comparison, other countries like the U.K. had developed large bodies of specialized inspectors to perform ongoing, detailed safety inspections of hazardous facilities. OSHA responded in part to the recommendation by creating a new National Emphasis Program for refineries; the program was considered very effective by OSHA leaders, uncovering many safety problems in refineries. Unfortunately, OSHA did not have adequate resources to continue the program for more than a temporary period.

The EPA has also lacked the dedicated resources to conduct extensive enforcement of RMP program requirements. When this Committee conducted oversight of the program in 2007, the EPA told the late Senator Lautenberg that the total RMP-related fines collected for the entire country over nearly a four-year period (from fiscal year 2004-2007) were just over $3.5 million,[8] a modest sum for a program that covers over 12,000 facilities.

The CSB believes there are a number of serious challenges for improving industrial process safety in the U.S. As noted above, both OSHA and EPA process safety standards rely heavily upon list-based approaches for determining which facilities and companies have to comply with the most rigorous requirements. This concept of a hazardous chemical list was largely borrowed from environmental statutes of the 1970's and 1980's. However, process safety experts generally recognize that process hazards are a function of chemistry itself, and it makes little sense to assert that the overall risks from chemical processing and handling can be adequately captured using small lists of chemicals. Time and again the CSB has found large chemical hazards – capable of causing major disasters – residing in facilities that have largely escaped regulatory scrutiny. These facilities – of which West Fertilizer is but one example – fall outside the scope of existing regulatory standards, which were developed in the 1990's and have seen few updates since then. All too often, a tragedy like the one at West suddenly exposes the hazards of a chemical or process that had somehow been overlooked.

The effects of these regulatory and enforcement challenges are evident in the accident rates for U.S. refineries and petrochemical sites. In 2008, a leading reinsurance company, Swiss Re, told the CSB and federal regulatory agencies that property losses from U.S. refinery accidents were occurring at approximately four times the rate of the rest of the world. In a follow-up briefing, Swiss Re officials asserted the gap between refinery safety performance in the U.S. and in the rest of the world was continuing to widen. Many developed nations have adopted a different approach for controlling major process hazards. For example, nations in Europe and elsewhere have implemented a "safety case" regime, that requires hazardous facilities to continuously meet higher standards and reduce risk. Companies work directly with the regulator to identify the most appropriate safety standards from around the world, which they then are required to follow as a condition of operating. The focus is on preventing accidents in highly complex, technological systems rather than post-accident punishment.

Implementing an effective regulatory regime such as the safety case, with the ability to manage and regulate high hazard industries and prevent serious accidents, requires a number of interdependent features. First, the regulatory regime must be truly goal-setting in nature; another term for this is a performance-based regulatory regime. This approach provides industry the opportunity to tailor the regulations to its specific facilities with the goal of continuous risk reduction and incident prevention. The safety case regime also imposes a general duty on industry to reduce all risks in its operations to as low as reasonably practicable (ALARP). Such an approach places the impetus on industry to evolve with current best safety practices, wherever they have been developed anywhere in the world, to ensure that process hazards have been adequately identified, evaluated, and controlled. Furthermore, this regime requires industry to utilize leading and lagging indicators to drive risks involved in major hazard facilities to as low as reasonably practicable. Finally, for effective implementation, this type of regime requires an independent, competent, and well-funded regulator. Experience and competence in technical areas such as chemical engineering, human factors, and process safety management are necessary to provide effective auditing and regulatory oversight for prevention. In a recent federal OSHA forum on reforming process safety regulations, noted safety expert Andrew Hopkins pointed out that all of these elements are essential for an effective major accident prevention regime. Dr. Hopkins emphasized that the whole package of the safety case system needs to be introduced to make it work, including a competent, well-funded regulator.[9]

The CSB has begun to examine these alternative regulatory systems in the context of investigating the recent Chevron refinery fire in California and the Deepwater Horizon blowout in the Gulf. This April, the CSB issued its interim report on the Chevron refinery fire, which sent over 15,000 Richmond residents to the hospital in August 2012. California legislators have responded proactively to the accident and to the CSB's recent findings and recommendations. A bill now before the California governor for signature would effectively triple the number of dedicated process safety inspectors in the state. This expansion will be funded by fees collected from the industry, and will not significantly burden taxpayers. And state legislators as well as leaders from Contra Costa County, where the refinery is located, have been working to implement other CSB recommendations for safer equipment designs and materials, reporting of process safety indicators, and improved maintenance procedures. California's actions should be closely examined, we believe, as a potential model for other states and the federal government to follow.

Thank you again, Chairman Boxer and Ranking Member Vitter, for the opportunity to testify today.

Photos from the June 13, 2013,
Williams Olefins Incident
Geismar, LA

CSB

End Notes

[1] Within the past three weeks, the ATF has begun producing records and evidence from its investigation to the CSB. The ATF released the remains of the West site from its control back to the company on May 24.

[2] Shah, K.D.; Roberts, A.G.; “Safety Considerations in the Processing, Handling, and Storage of Ammonium Nitrate;” In Keleti, C. (ed.); Nitric Acid and Fertilizer Nitrates; New York: Marcel Dekker Inc., 1985.

[3] NFPA 400 refers users to a 1953 publication by the U.S. Bureau of Mines for information on the explosive properties of AN.

[4] Calcium ammonium nitrate (CAN) must still be protected from contamination with other chemicals that can re-sensitize it to detonation. See Popovici Ipochim, N.N.; Icechim, M.M.; “Other Ammonium Nitrate Fertilizers;” In Keleti, C. (ed.); Nitric Acid and Fertilizer Nitrates; New York: Marcel Dekker Inc., 1985.

[5] Ibid.

[6] U.K. Health and Safety Executive; “Storing and Handling Ammonium Nitrate;” Available from http://www.hse.gov.uk/pubns/indg230.pdf

[7] http://www.cfindustries.com/pdf/Ammonium-Nitrate-Amtrate-MSDS.pdf

[8] Christopher P. Bliley, Associate Administrator, EPA; Letter to Senator Barbara Boxer, Chairman, Committee on Environment and Public Works, August 22, 2007.

[9] OSHA Expert Forum on the Use of Performance-Based Regulatory Models in the U.S. Oil and Gas Industry, Offshore and Onshore; Texas City, Texas; September 20, 2012.

In: Ammonium Nitrate Facilities
Editor: Allegra Ward
ISBN: 978-1-63321-716-4

Chapter 5

TESTIMONY OF BARRY N. BREEN, PRINCIPAL DEPUTY ASSISTANT ADMINSTRATOR, OFFICE OF SOLID WASTE AND EMERGENCY RESPONSE, U.S. ENVIRONMENTAL PROTECTION AGENCY. HEARING ON "OVERSIGHT OF FEDERAL RISK MANAGEMENT AND EMERGENCY PLANNING PROGRAMS TO PREVENT AND ADDRESS CHEMICAL THREATS, INCLUDING THE EVENTS LEADING UP TO THE EXPLOSIONS IN WEST, TX AND GEISMAR, LA"*

Good morning Chairman Boxer and members of the Committee, I am Barry Breen, Principal Deputy Assistant Administrator for the U.S Environmental Protection Agency's Office of Solid Waste and Emergency Response. Thank you for the opportunity to testify today on the EPA's Risk Management Program and emergency planning and community right-to-know issues.

---

* This is an edited, reformatted and augmented version of testimony presented June 27, 2013 before the Senate Committee on Environment and Public Works.

WEST, TEXAS FACILITY AND GEISMAR, LA INCIDENTS

On April 17, 2013, a fire and explosion occurred at the West Fertilizer plant in the town of West, Texas, causing multiple injuries and fatalities. The explosion shock wave caused multiple fires within a six block radius. The EPA responded as part of a multi-agency effort, including the U.S. Chemical Safety Board (CSB), the Federal Bureau of Alcohol Tobacco and Firearms (ATF), the Texas Commission on Environmental Quality (TCEQ), and Texas Fire Marshal Office. As part of the EPA's role, the agency conducted air monitoring using both stationary sites and a mobile monitoring team in the neighborhoods west of the facility. The EPA monitored for airborne contaminants including volatile organic compounds, ammonia, carbon monoxide, and lower explosive limits of methane gas. The EPA also deployed emergency response personnel to the site of the explosion and fire at the Williams Olefin facility in Geismar, LA. The agency is conducting its post-accident assessment efforts in coordination with the other federal, state and local agencies for both incidents.

THE EMERGENCY PLANNING AND COMMUNITY RIGHT-TO-KNOW ACT

In response to the devastating chemical disaster in Bhopal, India in 1984, Congress passed the Emergency Planning and Community Right-to-Know Act (EPCRA) in 1986 to ensure that local communities have the authority they need to prevent, prepare for, and respond to chemical accidents. The EPCRA provisions help increase local planners, responders, and the public's knowledge and access to information on chemicals at individual facilities and risks associated with them. States and communities, working with facilities, can use the information to improve chemical safety and protect public health and the environment. The implementing regulations for emergency planning, emergency release notification, and the chemicals subject to these regulations are codified in 40 CFR part 355. The implementing regulations for community rightto-know reporting (or hazardous chemical reporting) are codified in 40 CFR part 370.

Subtitle A of EPCRA establishes the framework for local emergency planning. The Act requires that the EPA publish a list of extremely hazardous substances (EHSs). The EHS list was established by the EPA to identify chemical substances that could cause serious irreversible health effects from accidental releases {(See 40 CFR part 355 (52 FR 13378, April 22, 1987)}. The Agency was also directed to establish a threshold planning quantity (TPQ) for each extremely hazardous substance.

The purpose of the EHSs list is to focus initial efforts in the development of state and local contingency plans. Inclusion of a chemical on the EHSs list indicates a need for the community to undertake a program to investigate and evaluate the potential for accidental exposure associated with the production, storage or handling of the chemical at a particular site and develop a chemical emergency response plan around those risks.

Under EPCRA section 302, a facility that has an EHS on-site in excess of its TPQ must notify the State Emergency Response Commission (SERC) and Local Emergency Planning Committee (LEPC), as well as participate in local emergency planning activities. Under the Statute, the LEPC shall then develop a community emergency response plan. Emergency Response plans contain information that community officials can use at the time of a chemical accident.

The EPA and the National Oceanic and Atmospheric Administration (NOAA) have developed a system of software applications used widely by States and local emergency planning committees to plan for and respond to chemical emergencies. This system is called the Computer-Aided Management of Emergency Operations (CAMEO) and it was developed to assist front-line chemical emergency planners and responders. Emergency responders and planners use CAMEO to access, store, and evaluate information critical for developing emergency plans. In addition, CAMEO supports regulatory compliance by helping users meet the chemical inventory reporting requirements of EPCRA. The CAMEO system integrates a chemical database and a method to manage the data, an air dispersion model, and a mapping capability. All modules work interactively to share and display critical information in a timely fashion.

Subtitle B of EPCRA established community right-to know requirements in order to ensure information on chemicals in the community is provided to the public as well as emergency responders. Under ECPRA sections 311 and 312, facilities that have either (1) a hazardous chemical present at or above 10,000 pounds or (2) an EHS present at or above its TPQ or 500 pounds—whichever is the lesser, are required to submit an Emergency and Hazardous Chemical Inventory form (Tier II) and a Material Safety Data Sheet (MSDS) for that chemical to their SERC, LEPC and local fire department. A chemical is hazardous as defined under the Hazard Communication Standard (HCS) of the Occupational Safety and Health Act (OSHA). There is not a separate list of hazardous chemicals. If a facility is required by OSHA to develop and/or maintain a MSDS for that chemical and it is present at or above the threshold discussed above, it must be reported. Local fire departments receive this information and should use it to understand the chemical(s) present at facilities in their community and precautions they may need to take in responding to an accident at the facility.

Sections 311 and 312 of EPCRA make available to the local and state emergency planners information on other chemicals and facilities, beyond those identified under section 302, that they may wish to include in their emergency planning efforts. The EPA has specified in guidance that Tier II information under section 312 will provide specific information on the quantities and locations of hazardous chemicals. Thus, sections 311 and 312 provide information supportive of the emergency planning required under Subtitle A. The facilities identified as a result of that subtitle are only a "first cut" of the facilities and potential chemical hazards for which emergency planning may be necessary.

RISK MANAGEMENT PROGRAM

The Clean Air Act (CAA) 112(r) provisions build on the planning and preparedness groundwork laid by EPCRA. CAA 112(r) provides the authority for the EPA's Risk Management Program (RMP). RMP regulations apply to the owner or operator of a stationary source with more than a threshold quantity of a CAA section 112(r) regulated substance in a process. Section 112(r) chemicals and thresholds may overlap with chemicals listed under other rules, but are not identical to those on any other list. The section 112(r) list includes 63 flammable gases and liquids and 77 acutely toxic chemicals. To develop the list, several statutory factors were considered, including the severity of any acute adverse health effects associated with accidental releases of the substance, the likelihood of accidental releases of the substance, and the potential magnitude of human exposure to accidental releases of the substance. An accidental release is an unanticipated emission of a regulated substance or other extremely hazardous substance into the ambient air from a stationary source. Many of these substances are also included on the EPCRA extremely hazardous substance (EHS) list. The section 112(r) chemical list and corresponding thresholds for each chemical are published at 40 CFR 68.130. Under CAA section 112 (r), the EPA is required to review the list of chemicals every 5 years or by its own motion or by petition. The EPA also provides an ongoing review of new chemicals and hazards to see if any chemical warrants listing or delisting.

Under the RMP regulations, a covered facility is required to review the hazards associated with the covered substance, process and procedures, as well as develop an accident prevention program and an emergency response program. The "Hazard Review" must identify opportunities for equipment malfunction or human error that could in turn cause the accidental release of the covered substance, as well as safeguards to prevent the potential release, and steps to detect and monitor for a release. A facility's compliance with these requirements is documented in a Risk Management Plan that is submitted to the EPA. Covered facilities must implement the Plan and update them every 5 years or when certain changes occur. The goal of the EPA's Risk Management Program is to prevent accidental releases of substances to the air that can cause serious harm to the public and the environment from short-term exposures, and to mitigate the severity of releases that do occur. Approximately 12,800 facilities are currently covered under Risk Management Program regulations.

Under the CAA section 112(r) RMP facilities must submit a risk management plan which includes:

- Facility hazard assessments, including worst-case release and alternative release scenarios;
- Facility accident prevention activities, such as use of special safety equipment, employee safety training programs, and process hazards analyses conducted by the facility;
- Past chemical accidents at a facility; and
- Facility emergency response programs and plans.

Another key component of Section 112(r) of the Clean Air Act, is section 112(r)(1), which is the General Duty Clause. This provision requires owners and operators of any stationary sources producing, processing, handling or storing an RMP substance or any other extremely hazardous substance to identify hazards which may result from such releases using appropriate hazard assessment techniques, to design and maintain a safe facility taking such steps as are necessary to prevent releases, and to minimize the consequences of accidental releases which may occur. This requirement is all encompassing and is used proactively to prevent accidents when hazards are observed that could lead to a chemical accident, or after an accident, if a facility failed to properly carry out this statutory requirement. Under the General Duty, facilities are expected to comply with recognized and generally accepted good engineering practices.

Both EPCRA and the CAA section 112(r) Risk Management Program encourage communication between facilities and the surrounding communities about chemical safety and chemical risks. Regulatory requirements, by themselves, will not guarantee safety from chemical accidents. Those who are handling hazardous substances must take the responsibility and act to prevent, prepare for and respond to chemical emergencies. Information about hazards in a community will allow local emergency officials and the public to work with industry to prevent accidents.

CONCLUSION

The EPA will continue its efforts to help prevent chemical accidents and releases under the Risk Management Program. Strong chemical accident prevention, preparedness, and response programs rely upon effective partnerships with the public and all levels of government. We will continue our outreach efforts to stakeholders and work with our federal, state, and local partners to promote chemical safety, address chemical process safety issues, and explore opportunities for improving chemical safety.

In: Ammonium Nitrate Facilities
Editor: Allegra Ward

ISBN: 978-1-63321-716-4

Chapter 6

TESTIMONY OF PAUL ORUM, CONSULTANT, COALITION TO PREVENT CHEMICAL DISASTERS. HEARING ON "OVERSIGHT OF FEDERAL RISK MANAGEMENT AND EMERGENCY PLANNING PROGRAMS TO PREVENT AND ADDRESS CHEMICAL THREATS, INCLUDING THE EVENTS LEADING UP TO THE EXPLOSIONS IN WEST, TX AND GEISMAR, LA"*

My name is Paul Orum. I thank the committee for the opportunity to present views important to a broad coalition of environmental health, labor, and community organizations known as the Coalition to Prevent Chemical Disasters. My background for 25 years is government information policy regarding hazardous materials.

Recent deadly explosions in West, Texas and Geismar, La., among others, remind us of the need for more effective public protections from industrial chemicals in populated areas.

* This is an edited, reformatted and augmented version of testimony presented June 27, 2013 before the Senate Committee on Environment and Public Works.

- These recent incidents are hardly rare. The National Response Center recorded more than 11,000 oil and chemical spills in the last year alone.[1]
- The potential for large-scale incidents is ever present. A Congressional Research Service analysis indicates more than 470 facilities have vulnerability zones potentially affecting any of 100,000 or more people in the event of a worst-case toxic gas release.[2]

Similar scenarios repeat. The fire and explosion at West Fertilizer is reminiscent of an event in Kansas City, Missouri, at which a construction facility storing ammonium nitrate first caught fire and then exploded killing six firefighters after they had responded to the fire. That was November 29, 1988.

In general, the chemical safety landscape includes a lot of neglect, missed communication, static regulations, voluntary standards, and prosecution afterwards. There is not enough on prevention, technically competent inspections, community-wide awareness, producer responsibility, and safer alternatives. Regulations should not only control problems but also generate safer solutions. Accident prevention is ultimately more effective than response.

Risk management and emergency planning should be revised and updated in light of ongoing and recent plant explosions.

1) Risk management planning should include reactive chemicals like the ammonium nitrate that detonated at West Fertilizer. Where there is serious potential harm to the public, reactive chemical hazards should be included in Risk Management Plans (RMP) under the Clean Air Act, section 112(r). The Chemical Safety and Hazard Investigation Board has an open recommendation to EPA to this end:

Revise the Accidental Release Prevention Requirements, 40 CFR 68, to explicitly cover catastrophic reactive hazards that have the potential to seriously impact the public, including those resulting from self-reactive chemicals and combinations of chemicals and process-specific conditions. (Recommendation No. 2001-1-I-H-R3)

While the general duty clause of the clean air act presumably covers all facilities that hold extremely hazardous substances – including reactive substances that pose catastrophic hazards – the general duty does not explicitly cover important *proactive* elements of RMPs, such as the requirement to assess and communicate chemical hazards. Adding ammonium nitrate to the RMP program could have informed the owner of West Fertilizer, first responders, and the public about the magnitude of the danger, including off-site consequences, and might have prevented or reduced the tragic consequences of the explosion.

2) Management systems and controls do fail. Chemical facility owners and operators have a responsibility not only to understand their own chemical hazards, but also to understand less hazardous alternatives that are commercially available in their industry. EPA should require chemical facilities to review and include in RMPs available methods that prevent potential consequences of a worst-case incident. Such methods are often the most effective measures to protect workers at the site, emergency responders, and nearby populations.

Surveys show that the RMP process has prompted some companies to reduce or remove chemical hazards, one of the objectives of the program. The RMP process facilitates changes that companies may be considering for a variety of reasons, including safety, security, and other regulatory requirements.

- More than 554 drinking water and wastewater facilities converted from toxic inhalation hazard chemicals, removing dangers to more than 40 million Americans. (The 554 facilities are examples among other facilities that have converted to less hazardous operations.)[3]
- Facilities across some 20 industries already use options that do not pose the danger of a major toxic gas release, including bleach producers, water utilities, power plants, refineries, aluminum smelters, and many types of manufacturers.[4]
- Facilities that convert to safer operations may save money when all factors are considered, such as avoided costs of release control devices, liability insurance, regulatory compliance, personal protective equipment, site security, and emergency planning.[5]

These facilities typically substituted a less hazardous replacement chemical or process; used a chemical in a less hazardous form (such as less concentrated, or aqueous instead of gaseous); or adjusted the process design to minimize use or storage (such as generating the chemical on site as-needed without storage). These strategies are distinct from conventional risk management approaches such as containment, control, mitigation, or recovery of substances.

The House and Senate reports on the Clean Air Act Amendments of 1990 show that Congress viewed measures to remove avoidable chemical hazards as integral to the statutory goal of preventing accidental releases:

> Measures which entirely eliminate the presence of potential hazards (through substitution of less harmful substances or by minimizing the quantity of an extremely hazardous substances present at any one time), as opposed to those which merely provide additional containment, are the most preferred.[6]
>
> Hazard assessments...include a review of the efficacy of various release prevention and control measures, including process changes or substitution of materials. [7]

EPA took public comment on inherently safer approaches for facility design and operations when first implementing the RMP program.[8] Unfortunately the agency did not develop the approach at the time. As a result, covered facilities are not required to evaluate feasible chemical hazard reduction alternatives that may be the most effective safety measures. Basic prevention analysis elements such as the avoided costs and liabilities associated with alternate technologies are not standard elements of RMPs. Such elements are foundational to developing knowledge of solutions. They are among the elements that help make organizations intelligent about the advantages, costs, and feasibility of technology options.

In March 2012, EPA's National Environmental Justice Advisory Council urged the agency to prevent chemical disasters by more fully using its authorities to advance safer chemical processes under the Clean Air Act.[9] In July 2012, more than 50 organizations petitioned EPA to commence rulemaking under the Clean Air Act and to revise agency guidance for enforcement of the general duty clause.[10]

The EPA Administrator has authority under the Clean Air Act, section 112(r), to incorporate methods that prevent potential consequences into RMPs and should do so.

3) The explosion at West Fertilizer illustrates the importance of the Clean Air Act's general duty to operate safely. West Fertilizer was subject to an incomplete patchwork of chemical safety regulations regarding ammonium nitrate. The general duty clause holds firms responsible for understanding and managing their chemical hazards

regardless of the completeness of government actions to regulate those hazards. For example, the ammonium nitrate at West Fertilizer was not on the RMP list of substances and thresholds. The general duty is an important tool for not only enforcement but also prevention. EPA's implementation guidance for the general duty clause recognizes that removing chemical hazards can be an effective safety measure, but EPA should further develop the concept in this guidance. We strongly oppose restricting the general duty clause in ways that could hamper enforcement or prevention. We also oppose arbitrarily fragmenting federal authorities between safety and security. By Presidential directive, the U.S. EPA is the lead agency to oversee security at drinking water and wastewater facilities.[11]

4) EPCRA emergency planning notification is incomplete. The ammonium nitrate that exploded at West Fertilizer was not on the EPCRA section 302 list of substances that require emergency planning notification. EPCRA section 302 requires facilities that hold threshold amounts of listed chemicals to notify their State Emergency Response Commission (SERC) and designate a point of contact at the facility to participate in emergency planning. It should be acknowledged that local emergency response capacities are often starkly overmatched by the magnitude of chemical hazards, and that activity levels of Local Emergency Planning Committees (LEPC) vary widely. Too much is left to the mostly-volunteer LEPCs – states should have fee-based programs that support hazard reduction, inspections, and regular drills. Nonetheless, EPCRA 302 notifications are a starting point for local emergency planning. The EPA Administrator has responsibility to modify the EPCRA 302 list and should do so. While lists and thresholds will inevitably fall short – hence the need for a general duty to operate safely – EPA should revise the EPCRA 302 list to include common substances that are known emergency hazards. This process should include both proactive listing criteria and a review of substances involved in serious incidents reported to the National Response Center.
5) EPCRA inventory reporting is valuable but insufficient. Owners and operators of facilities that hold large amounts of hazardous chemicals have an obligation to clearly communicate chemical hazards to those who could be affected prior to an emergency. West Fertilizer did report ammonium nitrate to the Texas SERC under EPCRA section 312 (a Tier II report). Texas apparently maintains Tier II reports in an

electronic format, which is important. EPA should continue to support and promote free electronic information management tools such as Tier II Submit, RMP*Comp, and CAMEO. The EPA should also develop routine electronic access to EPCRA 312 Tier II data from each state through memoranda of understanding or other means (as should OSHA and DHS). EPA should also promote awareness of reporting and planning obligations among regulated facilities. However, simple awareness of chemicals on-site is not sufficient. Local emergency planners and responders need not only chemical inventories but also worst-case and planning-case scenarios (which are included in RMPs but not EPCRA Tier II reports). They also need regular information about the number and type of high-hazard shipments in all modes of transportation. Fee-based programs should support prevention, pre-fire planning, technically competent inspections, drills, and NFPA-compliant hazmat training – including clear reminders that evacuating may be the most prudent course of action.

6) Independent investigations are important. The Chemical Safety and Hazard Investigation Board, also established by the Clean Air Act 112(r), produces root cause investigations and safety recommendations after the most serious chemical accidents. These activities are important to the public because they provide credible information and focused recommendations for change. Barriers to effective investigations, such as site access and preservation, should be resolved.
7) Issues beyond EPCRA and Clean Air Act, 112(r): Schools and nursing homes shouldn't be in potential blast zones. It is not an easy problem. Communities may grow up around chemical facilities or vice versa, but they are too close together in many places. State and local planners could benefit from federal guidelines for substantial safe setback distances, based on a worst-case scenario, in order not to continue to compound the problem when siting new buildings. School buildings were badly damaged by the blast in West, Texas. School siting criteria should take into account proximity to hazardous chemical facilities. Recipients of federal construction funds for buildings that will be used by potentially vulnerable populations (such as head start schools, hospitals, or nursing homes) should be subject to oversight to prevent building in the near zone of potential harm. In addition, the agricultural chemicals security tax credit assists

agricultural distributors with conventional security measures such as fences and lights; it should assist facilities that want to move locally to safer locations.

8) Hazardous chemical operations shouldn't be underinsured. West Fertilizer reportedly carried only $1 million in liability insurance, a fraction of the estimated $100 million in property damage alone. Companies that hold large amounts of extremely hazardous substances should be required to maintain sufficient liability insurance to cover a worst-case chemical release. Such a requirement would provide a reasonable cost incentive for companies to develop and use feasible alternatives. In addition, common carrier obligations encourage widespread overuse of railcars for shipping and storing extremely hazardous substances. Railroads have sought to have shippers share liability risks associated with extremely hazardous substances (which they are required to carry) and to have shippers develop safer substitutes.[12]

   Sustained improvement in chemical hazard prevention, preparedness, and response is long term and involves a range of actions. Among the most immediate lessons from the West Fertilizer explosion are for EPA to make sure major recognized hazards are 1) included in the programs designed to address them, 2) subject to safer alternatives analysis by the companies that hold them, 3) covered by appropriate lists and thresholds, and by the general duty to operate safely.

Thank you again for the opportunity to testify. I would be glad to take any questions.

End Notes

[1] On-line search of National Response Center conducted June 20, 2013. NRC is the national point of contact for reporting oil and chemical spills.

[2] Congressional Research Service memorandum to Senator Frank R. Lautenberg, RMP Facilities in the United States, November 16, 2012.

[3] Center for American Progress, Leading Water Utilities Secure Their Chemicals, March 2010.

[4] Center for American Progress, Chemical Security 101: What You Don't Have Can't Leak, or Be Blown Up by Terrorists, November 2008.

[5] Center for American Progress, Preventing Toxic Terrorism: How Some Chemical Facilities Are Removing Danger to American Communities, April 2006.

[6] Senate Report on the Clean Air Act Amendments of 1990 Report # 101-228 (S-1630), page 209.

[7] House of Representatives, Clean Air Act of 1990: Conference Report to Accompany S-1630. Report #101-952 (October 26, 1990), page 349.

[8] 60 Federal Register 13526, March 13, 1995.

[9] National Environmental Justice Advisory Council letter to EPA Administrator Lisa P. Jackson, March 14, 2012.

[10] Petition to the Environmental Protection Agency to Exercise Authority Under Section 112(r) of the Clean Air Act to Prevent Chemical Facility Disasters Through the Use of Safer Chemical Processes, July 25, 2012.

[11] Homeland Security Presidential Directive/HSPD 7, Critical Infrastructure Identification, Prioritization, and Protection, December 17, 2003.

[12] Center for American Progress, Toxic Trains and the Terrorist Threat: How Water Utilities Can Get Chlorine Gas Off the Rails and Out of American Communities, April 2007.

In: Ammonium Nitrate Facilities
Editor: Allegra Ward

ISBN: 978-1-63321-716-4

Chapter 7

Testimony of Dr. M. Sam Mannan, Regents Professor and Director, Mary Kay O'Connor Process Safety Center, Texas A&M University. Hearing on "Oversight of Federal Risk Management and Emergency Planning Programs to Prevent and Address Chemical Threats, Including the Events Leading up to the Explosions in West, TX and Geismar, LA"*

Introduction

My name is M. Sam Mannan and I hold a BS, MS, and PhD in chemical engineering. I am a registered professional engineer in the states of Louisiana and Texas, and I am a certified safety professional. I am a Fellow of the American Institute of Chemical Engineers and a member of the American Society of Safety Engineers, the International Institute of Ammonia

* This is an edited, reformatted and augmented version of testimony presented June 27, 2013 before the Senate Committee on Environment and Public Works.

Refrigeration, and the National Fire Protection Association. I am Director of the Mary Kay O'Connor Process Safety Center, holder of the T. Michael O'Connor Chair I in Chemical Engineering, and Regents Professor of Chemical Engineering at Texas A&M University. The Center was established to memorialize Mary Kay O'Connor, a chemical engineer who, along with 22 others, died in a chemical plant explosion in 1989 in Houston, Texas. The Center mission is to lead the integration of process safety – through education, research, and service – into the education and practice of all individuals and organizations involved in chemical operations. The vision of the Center is to serve as the premier process safety resource for all stakeholders so that safety becomes second nature for managers, engineers, and workers as progress continues toward zero injuries and lost lives. The Center seeks to develop safer processes, equipment, procedures, and management strategies that will minimize losses in the process industry. My area of expertise within the chemical engineering discipline is process safety. I teach process safety engineering both at the undergraduate and graduate level. I also teach continuing education courses on process safety and other specialty process safety courses in the United States and overseas. My research and practice is primarily in the area of process safety and related subjects. The opinions presented in this document represent my personal position on these issues. These opinions are based on my education, experience, research and training.

Risk management and emergency planning programs to prevent and address chemical threats are of extreme importance for the protection of the workforce, public, and the environment. These programs are also of great importance for the US national economy and security. I applaud the US Congress for continuing to pay attention to such important issues, and I appreciate the opportunity to provide my opinions in this process.

BACKGROUND

Chemicals play a key role in today's high-tech world. The chemical industry is linked to every technologically advanced industry. Only a handful of the goods and services we enjoy on a daily basis would exist without essential chemical products. Chemicals are also a big part of the economy in Texas and many other states. For example, the Texas chemical industry alone provides more than 100,000 jobs, and the state's chemical products are shipped worldwide at a value of more than $20 billion dollars annually.

But the use of chemicals is a two-edged sword. Safe use creates a healthier economy and a higher standard of living. Unsafe use threatens our lives, our businesses and our environment. As the industry's sophistication increases, so does the need to work and live safely with chemicals. In order to accomplish this, many stakeholders must work together diligently and with persistent determination. A common theme that also must be present is competence at all levels with regard to knowledge and execution of responsibilities.

Today's hearings are an appropriate congressional response to the recent events in West, Texas, and Geismar, Louisiana. Both these events were tragic and our heart goes out to the affected people, neighborhoods and cities, and the local authorities. We must as a nation and individuals explore and investigate these incidents and do our best to prevent the recurrence of such incidents. The hearings are focused on federal oversight programs, and I will limit majority of my testimony to that topic. However, because of the nature of accident prevention and role of all stakeholders, I will at times touch on those issues as well. Also, at the Center we had one PhD researcher working on ammonium nitrate before the West, Texas incident happened, and since the West, Texas incident, we have had a team of five PhD researchers working under my guidance on researching this whole issue and associated topics. Therefore, much of my testimony and opinions are derived from looking at the aftermath of the West incident. Wherever possible, I have tried to include information and knowledge derived from the Geismar, Louisiana, incident and its aftermath. I must also state that much is still unknown about these incidents and as the root causes are identified and more definitive information becomes available, some of these conclusions and opinions may have to be revisited.

The West, Texas, Incident

On Wednesday, April 18, 2013, an initial fire exacerbated into an explosion at West Fertilizer in West, Texas, causing the death of 15 people and injuring more than 200. The blast wave completely destroyed the facility and also caused varying levels of damage to many buildings, businesses, and homes at significantly long distances from the plant. More than 50 homes, a 50- unit apartment building, a nursing home and four schools were in the impact zone. Of the 15 people who died, 12 were emergency responders, who were responding to the initial fire and trying to control and extinguish the fire when the catastrophic explosion occurred.

The Geismar, Louisiana, Incident

On Thursday, June 13, 2013, an explosion occurred at Williams Olefins in Geismar, Louisiana, causing the death of two people and injuring more than 70.[1] Residents from a nearby community (St. Gabriel) were instructed to shelter in place.[2] This facility handles toxic chemicals and there was a concern about the air quality; therefore, the Louisiana Department of Environmental Quality (DEQ) checked the air quality during the subsequent days.[3] On the same day of the incident, the National Oceanic and Atmospheric Administration's (NOAA) Scientific Support Coordinator (SSC) was contacted by the US Coast Guard (USCG) regarding the plant fire and explosion at the Williams facility. Currently the USCG is requesting weather and plume modeling from NOAA.[4] An official report of total damages caused by the explosion is not available yet.

Federal Oversight Programs for Risk Management and Emergency Planning and Lessons Learned from West, Texas, Incident

The West Fertilizer facility had a capacity to store 110,000 lbs of ammonia and 540,000 lbs of ammonium nitrate (Tier II reporting data from 2012). The discussion below provides a summary of different federal regulations the West facility was required to comply with and the known status of such compliance and the oversight role played by the respective federal agencies.

OSHA Regulations

The Occupational Safety and Health Administration (OSHA) has general and specific regulations that would apply to the use and possession of Ammonium Nitrate (AN). Appendix A provides a more detailed discussion on potential coverage/oversight of the West Fertilizer facility by OSHA regulations and the regulatory requirements.

The West facility was required to comply with specific OSHA regulations, including the *Hazard Communication Standard* (29 CFR 1910.1200) and *Explosives and Blasting Agents Standard* (29 CFR 1910.109). While it is not clear what the compliance status of the facility was at the time of the incident, it can be argued that compliance with these programs could have prevented or mitigated the incident.

Compliance with the *Explosives and Blasting Agents Standard* also has many measures that would have prevented or mitigated the incident. For example, the ammonium nitrate was stored in a warehouse, in very close proximity to the seed area. "Ammonium nitrate shall be in a separate building or shall be separated by approved type firewalls of not less than 1 hour fire-resistance rating from storage of organic..."[5] Seed is an organic and combustible material, which could propagate the fire to areas where ammonium nitrate was stored. Storage of ammonium nitrate at an adequate distance from the seed area might have helped in preventing the explosion. It is unknown – but unlikely – whether the warehouse had firewalls. Firewalls would have prevented ammonium nitrate from heating and reaching the onset temperature of decomposition. The warehouse construction material was wood, which is also a combustible material. Overall, from what is known, the storage of ammonium nitrate at West Fertilizer Company did not provide adequate measures to prevent overheating and propagation of fire, which eventually lead to the explosion. "Not more than 2,500 tons (2270 tonnes) of bagged ammonium nitrate shall be stored in a building or structure not equipped with an automatic sprinkler system."[6]

Proper training on the hazards of ammonium nitrate and knowledge about a potential violent decomposition might have allowed firefighters to take a different approach when responding to and fighting the initial fire.

General requirements include the "*General Duty Clause*" of the Occupational Safety and Health Act (P.L. 91-596, as amended) and an Emergency Action Plan (EAP) according to OSHA Standard 1910.38[7]. The "*General Duty Clause*" requires employers to provide employees with a workplace that is free from *"recognized hazards that are causing or are likely to cause death or serious physical harm"*[8].

The Risk Management Plan submitted by West Fertilizer Company to the US Environmental Protection Agency states that the company had an EAP[9]. However, the EAP is not publicly available.

The West facility was not covered by the Process Safety Management (PSM) standard (29 CFR 1910.119) even though it stored large quantities of anhydrous ammonia (a listed chemical under the PSM standard). OSHA proposed to exclude retail facilities, oil and gas well drilling and servicing operations and normally unmanned remote facilities from the [PSM] standard.[10] A brief summary of the PSM standard is provided in Appendix A.

It should be noted that hazard analyses done under the PSM standard would have likely suggested prevention and mitigation measures similar to those provided under OSHA 1910.109 and NFPA 400.

The most recent known OSHA inspection of the West site was conducted in 1985. A fine of $30 was levied attributed to inadequate anhydrous ammonia storage and failures in Personal Protective Equipment (PPE).

EPA Regulations

Similar to OSHA, EPA also has a general duty clause and specific regulations that apply to the West Fertilizer facility. Appendix B provides a more detailed discussion on potential coverage/oversight of West Fertilizer by EPA regulations and the regulatory requirements.

Under the Clean Air Act Section 112(r)(1), the *General Duty Clause* states: *"The owners and operators of stationary sources producing, processing, handling or storing such substances [i.e., a chemical in 40 CFR 68 or any other extremely hazardous substance] have a general duty [in the same manner and to the same extent as the general duty clause in the Occupational Safety and Health Act (OSHA)] to identify hazards which may result from (such) releases using appropriate hazard assessment techniques, to design and maintain a safe facility taking such steps as are necessary to prevent releases, and to minimize the consequences of accidental releases which do occur."*

The *General Duty Clause* applies to any stationary source producing, processing, handling, or storing regulated substances or other extremely hazardous substances. "Other extremely hazardous substances" are any chemicals listed in 40 CFR 68, or any other chemicals, which may be considered extremely hazardous. Thus, it would seem the EPA has wide-ranging authority under the *General Duty Clause* to regulate West Fertilizer or other such facilities.

In addition to the EPA *General Duty Clause*, the following specific EPA regulations also apply to West Fertilizer:

- EPA's Risk Management Program (RMP) Rule (40 CFR 68) is intended to prevent and mitigate accidental releases of listed toxic and flammable substances. Requirements under the RMP rule include development of a hazard assessment, a prevention program, and an emergency response program. West Fertilizer would be regulated under the Program 2 requirements of the RMP rule because of the storage quantities of ammonia. Ammonium nitrate in not a listed substance under this rule.

- A separate EPA program, known as Tier II, requires reporting of hazardous chemicals (ammonium nitrate is included) stored above certain quantities. Tier II reports are submitted to local fire departments and emergency planning and response groups to help them plan for and respond to chemical disasters. In Texas, the reports are collected by the Department of State Health Services. As mentioned earlier, 2012 Tier II reporting data indicate that West Fertilizer filed a Tier II report stating that it had a capacity to store 540,000 lbs of ammonium nitrate at the facility.[11]

It could be argued that if the West Fertilizer facility had been regulated under Section 311 and 312 of EPCRA, the employees, fire responders and the community would have been more aware of the hazards of ammonium nitrate and consequences thereof. However, that argument is contingent on other factors including the fact that there is an operational and effective local emergency planning committee (LEPC) and other federal, state, and local government coordination.

The West Fertilizer facility last submitted a Risk Management Plan under the EPA Risk Management rule in June 30, 2011. In 2006, the EPA fined West Fertilizer Company with $2,300 for not having a risk-management plan that was up to federal standards.[12,13]

DHS Regulations

Within DHS, two regulations apply to the West Fertilizer facility. Appendix C provides a more detailed discussion on potential coverage/oversight of West Fertilizer by DHS regulations and the regulatory requirements.

One of the DHS regulations that may have applied to West Fertilizer has not been finalized yet and comes under the congressional statute, ***Section 563, Subtitle J, Secure Handling of Ammonium Nitrate Public Law 110-161***. As implied, this regulation primarily deals with the control of purchase and sales of ammonium nitrate. The other DHS regulation that applies to West Fertilizer is the Chemical Facility Anti-Terrorism Standard (CFATS).

It has been widely reported that West Fertilizer did not file a Top Screen report with DHS as required under the CFATS regulation. The facility was not inspected by DHS for compliance with the CFATS requirements, given its anticipated tier that may not have happened as of today.

DOT Regulations

West Fertilizer was covered by DOT regulations. Please see Appendix D for more details on the regulatory requirements for DOT.

All DOT requirements for ammonium nitrate are with regard to safe transportation. The last known inspection of the West Fertilizer site was conducted by DOT on September 23, 2011. The inspection resulted in a fine of $5,250 with a total of 2 violations; illegible data on ASME placards and/or missing flammable gas placards (front and/or rear) and no security plan.[14,15,16] All the penalties/fines were with regard to anhydrous ammonia.

ATF Regulations

Appendix E provides a more detailed description of the ATF regulatory requirements pertaining to ammonium nitrate. In summary, ATF regulations do not apply to ammonium nitrate used as fertilizer. However, ATF has embarked on several collaborative programs with industry organizations to improve security and safety at all ammonium nitrate facilities.

Conclusions and Recommendations

The incidents at West, Texas, and Geismar, Louisiana, are tragedies that could and should have been avoided. However, as I have stated before, this requires continued and committed efforts by all stakeholders. We in the academic community have embarked on some ground-breaking initiatives, but I will be the first one to admit that we have not done enough and we need to do more. So, with that caveat, please understand that when I criticize other stakeholders, I am happy to take criticism myself as well.

So, what should we do in the aftermath of the incidents in West, Texas, and Geismar, Louisiana? Clearly, as I have stated before, all stakeholders need to look at what they can do. However, this hearing is about the federal oversight programs on risk management and emergency planning. So, my conclusions and recommendations are primarily focused on that aspect.

1. Establishment of a national chemical incident surveillance system for process safety incidents. There is presently no reliable means for evaluating the performance of industry in limiting the number and severity of accidental chemical releases. There is also limited data with which to prioritize efforts to reduce the risks associated with such releases. Without this information, there are no means to measure the effectiveness of present programs or to guide future efforts.
2. Development of incident databases and lessons learned. This knowledge base could then be used to improve planning, response capability, and infrastructure changes. Recent experience in this regard is the improvement in planning and response for hurricane Rita from lessons learned from hurricane Katrina.
3. As a nation, we need to understand if regulations are doing what we intend them to do. To do that, we must understand the issues and to what agency to turn to find a solution. I strongly urge the US Congress to mandate a risk-based study to determine the hazards/risks and develop a regulatory map of hazardous materials oversight. This study should take into consideration types of facilities, their locations, chemicals involved and their quantities in order to determine what agencies do or do not regulate these facilities.
4. All federal agencies with responsibility to regulate safety/risk and associated issues should be required to conduct a primary screening to determine their regulatory landscape. Inter-agency training and briefings with regard to what each agency is covering and how they are enforced would also be beneficial.
5. Once the regulatory landscape is determined in item (4) above, each federal agency should be charged with developing a plan and schedule for ensuring compliance through regular inspections.
6. Inspections can only yield positive results when an adequate number of qualified, trained and competent inspectors is available. Clearly, in these days of budget restrictions, hiring and training hundreds or thousands more inspectors is going to be a challenge at least and at worst impossible. A cost-effective and viable alternative is third-party certified audits and inspections mentioned in item (7) below.
7. Congress should consider directing federal agencies to create verifiable and certified third-party auditing and inspection systems. This approach has worked for ISO-9000 certifications and other programs. There are market-based approaches through which this

regime can be implemented without causing a major burden on the regulatory authority or the regulated community. For example, refer to the studies done by the University of Pennsylvania's Risk Management and Decision Processes Center regarding third-party audits and inspections for EPA's Risk Management Program17 and Environmental Programs[18].

8. I believe that EPCRA Sections 301-303 provide a systematic framework for coordination of hazard information, prevention programs, and emergency planning and response involving the federal government, state emergency response commissions (SERC) and the local emergency planning committees (LEPC). However, because of a lack of systematic funding and operational capability, most LEPC's are dysfunctional or exist in name only. Some further examination into better communication between the federal and state partners is needed. I urge Congress to look into ways to solve this problem and utilize the LEPC framework in an effective manner.
9. The fact that a nursing home, schools, residential neighborhoods, and other public facilities were so near the blast zone in the West Fertilizer incident raises questions about zoning and land-use planning. I urge the US Congress to look into ways to encourage states and local governments to improve and enforce risk-based zoning and land-use planning.

Summary

I applaud the US Congress for providing leadership in this important area of risk management and emergency planning programs to prevent and address chemical threats. We have made a lot of progress in moving forward to overcome the challenges we face in using chemicals to improve our lives without hurting the industry employees, the public, and the environment. We all can agree that chemicals do improve our lives but we also can agree that they can hurt us too and I as have often said, if we do not do the right things, they can make us extinct as well. This is a serious matter and I am pleased that people at the highest level of government are involved at looking at this matter.

I am encouraged by the leadership of Congress and by continued efforts to seek expertise and opinion from all stakeholders.

APPENDIX A. POTENTIAL COVERAGE/OVERSIGHT OF WEST FERTILIZER BY OSHA REGULATIONS

The Occupational Safety and Health Administration (OSHA) has general and specific regulations that would apply to the use and possession of Ammonium Nitrate (AN). General requirements include the "General Duty Clause" of the Occupational Safety and Health Act (P.L. 91-596, as amended) and an Emergency Action Plan (EAP) according to OSHA Standard 1910.38[19]. The "General Duty Clause" requires employers to provide employees with a workplace that is free from *"recognized hazards that are causing or are likely to cause death or serious physical harm"*[20]. The Emergency Action Plan must have, at minimum, the following elements[21]:

- procedures for reporting a fire or other emergency;
- procedures for evacuation;
- procedures to be followed by employees who remain to operate parts of the facilities before evacuating;
- procedures to account for all employees after evacuation;
- procedures for employees performing rescue or medical duties; and
- names and job titles of persons who may be contacted by employees to provide information to employees about the EAP

The Risk Management Plan submitted by West Fertilizer Company to the US Environmental Protection Agency states that the company had an EAP[22]. However, the EAP is not publicly available.

Other specific regulations from OSHA that might potentially cover operations at the West, Texas facility include the following:

29 CFR 1910.109: Explosives and Blasting Agents[23]

Brief Summary of Regulation

This standard regulates the storage, use and transportation of explosives and blasting agents, including mixtures of fuel and oxidizers, *e.g.*, mixtures that might contain ammonium nitrate. Following is the definition of a blasting agent, according to OSHA Standard 1910.109:

1910.109(a)(1) – "Blasting agent - any material or mixture, consisting of a *fuel and oxidizer*, intended for blasting, not otherwise classified as an explosive and in which none of the ingredients are classified as an explosive, provided that the finished product, as mixed and packaged for use or shipment, cannot be detonated by means of a No. 8 test blasting cap when unconfined."

Section **1910.109(g)**, which addresses "blasting agents", specifically makes reference to ammonium nitrate handling in mixing facilities, and it provides recommended separation distances of ammonium nitrate and blasting agents from explosives or blasting agents.

In addition, section **1910.109(i)** provides specific requirements for the storage of ammonium nitrate – but does not apply to transportation. The following paragraphs are taken from OSHA Standard 1910.109:

1910.109(i)(1)(i)(a) – "Except as provided in paragraph (i)(1)(i)(d) of this paragraph applies to the storage of ammonium nitrate in the form of crystals, flakes, grains, or prills including fertilizer grade, dynamite grade, nitrous oxide grade, technical grade, and other mixtures containing 60 percent or more ammonium nitrate by weight but does not apply to blasting agents."

1910.109(i)(1)(ii)(b) – "The standards for ammonium nitrate (nitrous oxide grade) are those found in the "Specifications, Properties, and Recommendations for Packaging, Transportation, Storage, and Use of Ammonium Nitrate", available from the Compressed Gas Association, Inc., which is incorporated by reference as specified in Sec. 1910.6".

Compliance Requirements

1910.109(i)(2)(i) – "This paragraph applies to all persons storing, having, or keeping ammonium nitrate, and to the owner or lessee of any building, premises, or structure in which ammonium nitrate is stored in quantities of 1,000 pounds or more."

1910.109(i)(2)(ii) – "Approval of large quantity storage shall be subject to due consideration of the fire and explosion hazards, including exposure to toxic vapors from burning or decomposing ammonium nitrate."

Some of the specific requirements for the storage of ammonium nitrate, among others, which West Fertilizer Company should have complied with are the following:

1910.109(i)(2)(iii)(a) – "... Storage buildings shall not be over one story in height."

1910.109(i)(2)(iii)(b) – "Storage buildings shall have adequate ventilation or be of a construction that will be self-ventilating in the event of fire."

1910.109(i)(2)(iii)(c) – "The wall on the exposed side of a storage building within 50 feet of a combustible building, forest, piles of combustible materials and similar exposure hazards shall be of fire-resistive construction…"

1910.109(i)(2)(iii)(e) – "The continued use of an existing storage building or structure not in strict conformity with this paragraph may be approved in cases where such continued use will not constitute a hazard to life."

1910.109(i)(2)(iii)(f) – "Buildings and structures shall be dry and free from water seepage through the roof, walls, and floors."

1910.109(i)(4)(i)(a) – "Warehouses shall have adequate ventilation or be capable of adequate ventilation in case of fire."

1910.109(i)(7)(ii)(b) – "Water supplies and fire hydrants shall be available in accordance with recognized good practices."

Some of the requirements are summarized in the following table:

Description	OSHA 1910.109 Requirement
Piles size	H: 20 ft (6.1 m) W: 20 ft (6.1 m) L: 50 ft (15.2 m)
Piles – walls distance	30 inches (0.762 m)
Pile – roof distance	36 inches (0.91 m)
Pile – pile distance	3 ft (0.91 m)
Storage buildings requirements	The **wall** on the exposed side of a storage building within 50 ft of a combustible building = **fire resistant**
Contaminants	Include, but it is not limited to animal fats, baled cotton, baled rags, baled scrap paper, bleaching powder, burlap or cotton bags, caustic soda, coal, coke, charcoal, cork, camphor, excelsior, fibers of any kind, fish oils, fish meal, foam rubber, hay, lubricating oil, linseed oil, or other oxidizable or drying oils, naphthalene, oakum, oiled clothing, oiled paper, oiled textiles, paint, straw, sawdust, wood shavings, or vegetable oils.

Standard 1910.1200: Hazard Communication[24]

Brief Summary of Regulation

The goal of this standard is to ensure that employers provide employees adequate information about the hazards of all substances handled at the facility. West Fertilizer Company was covered under this regulation, according to the following paragraph taken from the standard:

> **1910.1200(b)(2)** – "This section applies to any chemical which is known to be present in the workplace in such a manner that employees may be exposed under normal conditions of use or in a foreseeable emergency."

Compliance Requirements

The West facility was required to comply with the following requirements (among others) under OSHA's hazard communication standard:

> **1910.1200(a)(2)** – "Classifying the potential hazards of chemicals and communicating information concerning hazards and appropriate protective measures to employees, may include, for example, but is not limited to, provisions for
>
> - developing and maintaining a written hazard communication program for the workplace,
> - including lists of hazardous chemicals present;
> - labeling of containers of chemicals in the workplace, as well as of containers of chemicals being shipped to other workplaces;
> - preparation and distribution of safety data sheets to employees and downstream employers; and
> - development and implementation of employee training programs regarding hazards of chemicals and protective measures."
>
> **1910.1200(d)(1)** – "Chemical manufacturers and importers shall evaluate chemicals produced in their workplaces or imported by them to classify the chemicals in accordance with this section. For each chemical, the chemical manufacturer or importer shall determine the hazard classes, and, where appropriate, the category of each class that apply to the chemical being classified. Employers are not required to classify chemicals unless they choose not to rely on the classification performed by the chemical manufacturer or importer for the chemical to satisfy this requirement."

The written hazard communication program should include the following:

1910.1200(e)(1) – "…at least describes how the criteria specified in paragraphs (f), (g), and (h) of this section for labels and other forms of warning, safety data sheets, and employee information and training will be met, and which also includes the following:

- A list of the hazardous chemicals known to be present using a product identifier that is referenced on the appropriate safety data sheet (the list may be compiled for the workplace as a whole or for individual work areas); and,
- The methods the employer will use to inform employees of the hazards of non-routine tasks (for example, the cleaning of reactor vessels), and the hazards associated with chemicals contained in unlabeled pipes in their work areas."

Standard 1910.119: Process Safety Management of Highly Hazardous Chemicals[25]

Brief Summary of Regulation

The Process Safety Management (PSM) standard *"contains requirements for preventing or minimizing the consequences of catastrophic releases of toxic, reactive, flammable, or explosive chemicals"* [26]. Even though West Fertilizer stored ammonia in excess of the threshold specified for ammonia in the PSM standard, this regulation did not apply to West Fertilizer Company, because of the exemption granted to retail facilities.

Compliance Requirements for Facilities Covered by the PSM Standard

Companies covered under the PSM standard must develop and implement a program covering the following 14 elements:

- Employee Participation
- Process Safety Information
- Process Hazard Analysis
- Operating Procedures
- Training
- Contractor Safety
- Pre-Startup Safety Review
- Mechanical Integrity
- Hot Work Program

- Management of Change
- Incident Investigation
- Emergency Planning and Response
- Compliance Audits
- Trade Secrets

APPENDIX B. POTENTIAL COVERAGE/OVERSIGHT OF WEST FERTILIZER BY EPA REGULATIONS

The US Environmental Protection Agency (EPA) is authorized to regulate production, distribution, storage, and release of most chemicals in commerce. The Emergency Planning and Community Right to Know Act (EPCRA) and Section 112(r) of the Clean Air Act (CAA) directly address the potential risks from facilities holding chemical hazards[27]. *Both EPCRA and the CAA section 112(r) Risk Management Program encourage communication between facilities and the surrounding communities about chemical safety and chemical risks*[28].

The Emergency Planning and Community Right-to-Know Act (EPCRA)

EPCRA has four major provisions[29]*:*

- Emergency planning (sections 301-303),Office of Emergency Management Factsheet EPCRA September 2012
- Emergency release notification (section 304),
- Hazardous chemical storage reporting requirements (sections 311-312), and
- Toxic chemical release inventory (section 313).

EPCRA, **Section 311**, requires owners or operators of local facilities covered by the Occupational Safety and Health Act to submit a material safety data sheet (MSDS) for each "**hazardous chemical**," or a list of such chemicals, to the SERC, the LEPC, and the local fire department.

EPCRA, **Section 312**, requires the same employers to submit annually an emergency and hazardous chemical inventory form to the SERC, LEPC, and local fire department. These forms must provide estimates of:

- Maximum amount of the chemicals present at the facility at any time during the preceding year
- Average daily amount of chemicals present
- General location of the chemicals in the facility

The West Fertilizer facility was exempt from the EPCRA requirements because of exemptions granted to retail fertilizer facilities. EPCRA Section 311(e)(5) excludes certain substances, including "fertilizer held for sale by a retailer to the ultimate customer."[30,31]

Section 112(r) of the Clean Air Act[32]

Background

- The Act requires EPA to promulgate an initial list of at least 100 substances that, in the event of an accidental release[33], are known to cause or may reasonably be anticipated to cause death, injury, or serious adverse effects to human health or the environment[34].
- In developing this list, EPA was required to consider, but was not limited to, the list of extremely hazardous substances (EHSs) promulgated under EPCRA (SARA Title III) section 302. EPA did not propose to adopt the entire EHS list because it includes a number of **solids and non-volatile liquids** for which an effect beyond the fenceline in the event of an accidental release is expected to be less likely than for gaseous or volatile liquids[35].
- Congress listed the following 16 substances to be included in the initial list (Chlorine, **ammonia and anhydrous ammonia**, methyl chloride, ethylene oxide, vinyl chloride, methyl isocyanate, hydrogen cyanide, hydrogen sulfide, toluene diisocyanate, phosgene, bromine, anhydrous hydrogen chloride, hydrogen fluoride, anhydrous sulfur dioxide, and sulfur trioxide).
- Explosive materials (Division 1.1. under DOT classification) were initially included in the list of highly hazardous materials when the EPA regulation was developed. However, explosive materials were delisted[36] in 1998 with the proviso that ATF covered all the aspects

that are necessary under RMP, except for public disclosure[37]. The industry voluntarily agreed to make that public disclosure that makes it equivalent to RMP.

Table B-1. Summary of categories and thresholds of extremely hazardous materials[38]

Categories	Requirements	Threshold quantities (lb)*
77 Toxic substances	*Acute toxicity:* • Inhalation: LC50 = 0.5 mg/L or • Dermal: LD50 = 50 mg/kg of body weight, or • Oral: LD50 = 25 mg/kg of body weight *Vapor pressure* >10 mmHg *Accident history*	500 – 20,000
63 Flammable substances	FP < 73 °F (22.8 °C) BP < 100 °F (37.8 °C)	10,000

* Substances in mixtures would be exempted from the threshold determination if they represent less than one percent of the mixture by weight. (EPA List of Regulated Substances is found in reference 39)[39].

The West Fertilizer facility was covered under Program 2 of the EPA Risk Management Program because of ammonia. However, ammonium nitrate is not included in the covered list and West Fertilizer would not have had to report any analysis or calculations regarding ammonium nitrate in their submissions to EPA.

Table B-1 shows a summary of the criteria used by EPA for determining extremely hazardous materials and the corresponding thresholds to be covered under the RMP rule. Based Table B-1, ammonium nitrate is not covered by the RMP rule because ammonium nitrate does not meet the requirements to be considered as toxic or flammable.

- **Listing criteria:**
    - **Toxicity:** Listed toxic substances are expected to rapidly become airborne, thus human exposure by the inhalation route is of primary concern. The listing criteria established for toxic substances considers not only acute toxicity, but also physical/chemical properties (physical state, vapor pressure), and accident history.

The acute toxicity criteria:
(a) Inhalation LC50 0.5 milligrams per liter of air (for exposure time 8 hours), or
(b) Dermal LD50 50 milligrams per kilogram of body weight, or
(c) Oral LD50 25 milligrams per kilogram of body weight.
Vapor pressure cut off:
Initially, a vapor pressure criterion of **0.5 mm Hg** was used as a baseline, based on the vapor pressure of toluene diisocyanate, a substance mandated for the initial list by Congress. However, EPA considered that this low vapor pressure level may lead to an overly conservative listing of chemicals that pose a relatively lower potential for air releases. Then, **EPA decided to set the vapor pressure criterion at the higher level of 10 mm Hg**. Substances with pressures above 10 mm Hg are likely to be volatilized and released, even after a timely facility response occurs, potentially causing off-site impacts.
Accident history:
Substances that "are known to cause ... death, injury, or serious adverse effects on human health or the environment" may be included on the list under section 112(r)(3).

- **Flammable gases and volatile flammable liquids:** Based on the flash point (FP) and boiling point (BP) criteria used by NFPA. Based on both accident reports and modeling results, EPA considered that flammable substances that meet the listing criteria, in quantities above the threshold quantity of 10,000 lb, could present a hazard to the public from a vapor cloud explosion. OSHA's PSM Standard provides an exemption for flammable liquids kept in atmospheric tanks below their normal boiling point. Unlike OSHA, EPA considers these substances to be intrinsically hazardous, regardless of conditions of storage, and, therefore, no exemption is provided in those cases.

Requirements if ammonium nitrate were covered by CAA 112

EPA defined three “program levels” to ensure that individual chemical processes are subject to appropriate requirements based on the size of the process and the associated risks[40].

- **Program 1 eligibility (provided in section § 68.10)[41].**
  1. For the five years prior to the submission of an RMP, the process has not had an accidental release of a regulated substance where exposure to the substance, its reaction products, overpressure generated by an explosion involving the substance, or radiant heat generated by a fire involving the substance led to any of the following offsite: (i) Death; (ii) Injury; or (iii) Response or restoration activities for an exposure of an environmental receptor.
  2. The distance to a toxic or flammable endpoint for a worst-case release assessment conducted under Subpart B and § 68.25 is **less than the distance to any public receptor**, as defined in §68.30.
  3. Emergency response procedures have been coordinated between the stationary source and local emergency planning and response organizations.
- **Program 1 requirements (provided in section § 68.12):**
  1. Analyze the worst-case release scenario for the process(es), as provided in § 68.25; document that the nearest public receptor is beyond the distance to a toxic or flammable endpoint defined in § 68.22(a); and submit in the RMP the worst-case release scenario as provided in § 68.165;
  2. Complete the five-year accident history for the process as provided in § 68.42 of this part and submit it in the RMP as provided in § 68.168;
  3. Ensure that response actions have been coordinated with local emergency planning and response agencies; and
  4. Certify in the RMP the following: Based on the criteria in 40 CFR 68.10, the distance to the specified endpoint for the worst-case accidental release scenario for the following process(es) is less than the distance to the nearest public receptor: Within the past five years, the process(es) has (have) had no accidental release that caused offsite impacts provided in the risk management program rule (40 CFR 68.10(b)(1)).
- **Program 2 eligibility (provided in section § 68.10):**

A covered process is subject to Program 2 requirements if it does not meet the eligibility requirements of program 1 and 3.

- **Program 2 requirements (provided in section § 68.12):**
  1. Develop and implement a management system as provided in § 68.15;
  2. Conduct a hazard assessment as provided in Sec.§ 68.20 through 68.42;
  3. Implement the Program 2 prevention steps provided in Sec. § 68.48 through 68.60 or implement the Program 3 prevention steps provided in Sec.§ 68.65 through 68.87;
- **Program 3 eligibility (provided in section § 68.10):**

  A covered process is subject to Program 3 if the process does not meet the requirements of program 1 of this section, and if either of the following conditions is met:
  1. The process is in NAICS code 32211, 32411, 32511, 325181, 325188, 325192, 325199, 325211, 325311, or 32532; **or**
  2. The process is subject to the OSHA process safety management standard, 29 CFR 1910.119.
- **Program 3 requirements (provided in section § 68.12):**
  1. Develop and implement a management system as provided in § 68.15;
  2. Conduct a hazard assessment as provided in Sec.§ 68.20 through 68.42;
  3. Implement the prevention requirements of Sec.§ 68.65 through 68.87;
  4. Develop and implement an emergency response program as provided in Sec.§ 68.90 to 68.95 of this part; and
  5. Submit as part of the RMP the data on prevention program elements for Program 3 processes as provided in § 68.175.

Figure B-1 can be used to identify the program level. In general, the requirements under the RMP rule include development of a hazard assessment, a prevention program, and an emergency response program.[42]

Based on the eligibility criteria, West Fertilizer Company would not be included in Program 1 because the distance to a toxic or flammable endpoint for a worst-case release assessment is greater than the distance to any public receptor. The West Fertilizer facility would not be included in Program 3 because the company NAICS code (42451-Facility grain and field bean merchant wholesalers) is not listed in the Program 3 eligibility requirements AND the West Fertilizer facility is excluded from the PSM program because of the retail exemption. Hence, the West Fertilizer facility would be covered by Program 2 of the EPA Risk Management Program, but only because of the storage of ammonia.

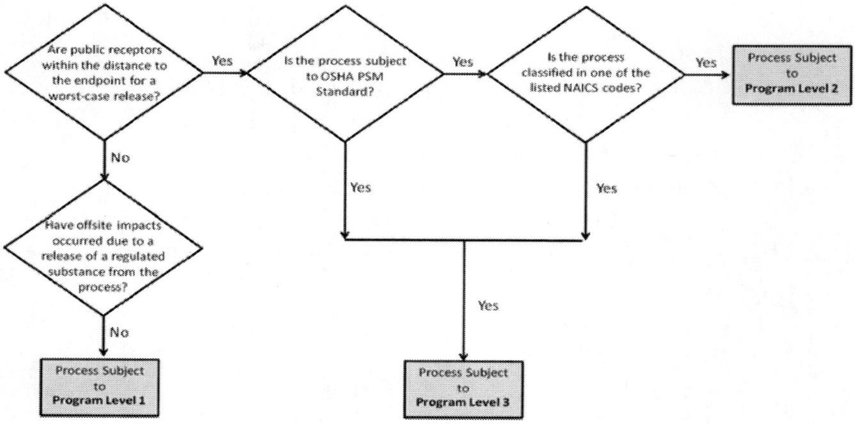

Figure B-1. Diagram of the decision rules on determining Program level.[43]

APPENDIX C. POTENTIAL COVERAGE/OVERSIGHT OF WEST FERTILIZER BY DHS REGULATIONS

With regard to ammonium nitrate, DHS regulations include the proposed rule regulating the control of the purchase and the sales of AN (Section 563) and the Chemical Facility Anti- Terrorism Standards (CFATS). To-date, DHS has not published the final rule mandated under the congressional statute (Section 563) summarized below. CFATS regulation is administered by DHS, and the requirements under CFATS are also discussed in this Appendix.

Section 563, Subtitle J, Secure Handling of Ammonium Nitrate Public Law 110-161

Section 563 of the 2008 Consolidated Appropriations Act, Subtitle J, Secure Handling of Ammonium Nitrate ("Section 563"), Public Law 110-161,[44] requires the Department of Homeland Security to "regulate the sale and transfer of ammonium nitrate by an ammonium nitrate facility ... to prevent the misappropriation or use of ammonium nitrate in an act of terrorism." [45]

''Subtitle J—Secure Handling of Ammonium Nitrate

SEC. 899A. DEFINITIONS.

SEC. 899B. REGULATION OF THE SALE AND TRANSFER OF AMMONIUM NITRATE.

SEC. 899C. INSPECTION AND AUDITING OF RECORDS. SEC. 899D. ADMINISTRATIVE PROVISIONS.

SEC. 899E. THEFT REPORTING REQUIREMENT. SEC. 899F. PROHIBITIONS AND PENALTY.

SEC. 899G. PROTECTION FROM CIVIL LIABILITY. SEC. 899H. PREEMPTION OF OTHER LAWS.

SEC. 899I. DEADLINES FOR REGULATIONS.

SEC. 899J. AUTHORIZATION OF APPROPRIATIONS.''

For example, SEC. 899B, states:

"SEC. 899B. REGULATION OF THE SALE AND TRANSFER OF AMMONIUM NITRATE.

(a) IN GENERAL.—The Secretary shall regulate the sale and transfer of ammonium nitrate by an ammonium nitrate facility in accordance with this subtitle to prevent the misappropriation or use of ammonium nitrate in an act of terrorism.

(b) AMMONIUM NITRATE MIXTURES.—Not later than 90 days after the date of the enactment of this subtitle, the Secretary, in consultation with the heads of appropriate Federal departments and agencies (including the Secretary of Agriculture), shall, after notice and an opportunity for comment, establish a threshold percentage for ammonium nitrate in a substance.

(c) REGISTRATION OF OWNERS OF AMMONIUM NITRATE FACILITIES.—

(1) REGISTRATION.—The Secretary shall establish a process by which any person that—

(A) owns an ammonium nitrate facility is required to register with the Department; and

(B) registers under subparagraph (A) is issued a registration number for purposes of this subtitle.

(2) REGISTRATION INFORMATION.—Any person applying to register under paragraph (1) shall submit to the Secretary—

(A) the name, address, and telephone number of each ammonium nitrate facility owned by that person;

(B) the name of the person designated by that personas the point of contact for each such facility, for purposes of this subtitle; and

(C) such other information as the Secretary may determine is appropriate.

(d) REGISTRATION OF AMMONIUM NITRATE PURCHASERS.—

(1) REGISTRATION.—The Secretary shall establish a process by which any person that—

(A) intends to be an ammonium nitrate purchaser is required to register with the Department; and

(B) registers under subparagraph (A) is issued a registration number for purposes of this subtitle.

(2) REGISTRATION INFORMATION.—Any person applying to register under paragraph (1) as an ammonium nitrate purchaser shall submit to the Secretary—

(A) the name, address, and telephone number of the applicant; and

(B) the intended use of ammonium nitrate to be purchased by the applicant.

(e) RECORDS.—

(1) MAINTE NANCE OF RECORDS.—The owner of an ammonium nitrate facility shall—

(A) maintain a record of each sale or transfer of ammonium nitrate, during the two- year period beginning on the date of that sale or transfer; and

(B) include in such record the information described in paragraph (2).

(2) SPECIFIC INFORMATION REQUIRED.—For each sale or transfer of ammonium nitrate, the owner of an ammonium nitrate facility shall—

(A) record the name, address, telephone number, and registration number issued under subsection (c) or (d) of each person that purchases ammonium nitrate, in a manner prescribed by the Secretary;

(B) if applicable, record the name, address, and telephone number of an agent acting on behalf of the person described in subparagraph (A), at the point of sale;

(C) record the date and quantity of ammonium nitrate sold or transferred; and

(D) verify the identity of the persons described in subparagraphs (A) and (B), as applicable, in accordance with a procedure established by the Secretary.

(3) PROTECTION OF INFORMATION .—In maintaining records in accordance with paragraph (1), the owner of an ammonium nitrate facility shall take reasonable actions to ensure the protection of the information included in such records.

(f) EXEMPTION FOR EXPLOSIVE PURPOSES .—The Secretary may exempt from this subtitle a person producing, selling, or purchasing ammonium nitrate exclusively for use in the production of an explosive under a license or permit issued under chapter 40 of title 18, United States Code.

(g) CONSULTATION .—In carrying out this section, the Secretary shall consult with the Secretary of Agriculture, States, and appropriate private sector entities, to ensure that the access of agricultural producers to ammonium nitrate is not unduly burdened.

(h) DATA CONFIDENTIALITY .—……

(i) REGISTRATION PROCEDURES AND CHECK OF TERRORIST SCREENING DATABASE .—

(1) REGISTRATION PROCEDURES .—

(A) GENERALLY .—The Secretary shall establish procedures to efficiently receive applications for registration numbers under this subtitle, conduct the checks required under paragraph (2), and promptly issue or deny a registration number.

(B) INITIAL SIX -MONTH REGISTRATION PERIOD .—The Secretary shall take steps to maximize the number of registration applications that are submitted and processed during the six-month period described in section 899F(e).

(2) CHECK OF TERRORIST SCREENING DATABASE .—

(A) CHECK REQUIRED .—The Secretary shall conduct a check of appropriate identifying information of any person seeking to register with the Department under subsection (c) or (d) against identifying information that appears in the terrorist screening database of the Department."

Chemical Facility Anti-Terrorism Standards (CFATS)

CFATS addresses hundreds of chemicals, including ammonium nitrate, and is directed at the security of high-risk facilities. DHS stated in the CFATS interim final rule that "if a retail establishment does exceed any of these [screening threshold quantities], *the retail establishment will have to complete the Top-Screen.*"[46]

The DHS lists 322 chemicals and screening threshold quantities for each chemical to determine the need to comply with CFATS[47]*. The DHS considers each chemical in the context of three threats: release; theft or diversion; and sabotage and contamination. The regulation lists two formulations of ammonium nitrate (one used as a blasting agent, the other as fertilizer) as a chemical of interest and identifies them as release and theft or diversion threats.*

The screening threshold quantity differs depending on whether the ammonium nitrate is a blasting agent or fertilizer. Facilities having at least 5,000 lbs of AN (400 lbs, if packaged for transportation), as a blasting agent (ammonium nitrate with more than 0.2% combustible substances), or at least 2,000 lbs of transportable fertilizer (with nitrogen concentration of 23% or greater, or fertilizer mixture containing at least 33% of AN) are considered a high risk facility. Therefore, they should follow CFATS[48].

"Assignment of tiers is based on an assessment of the potential consequences of a successful attack on assets associated with chemicals of interest. The Department of Homeland Security uses information submitted by facilities through the Chemical Security Assessment Tool Top Screen and Security Vulnerability Assessment (SVA) processes to identify a facility's risk, which is a function of the potential impacts of an attack (consequences), the likelihood that an attack on the facility would be successful (vulnerabilities), and the likelihood that such an attack would occur at the facility (threat). All facilities that were individually requested by the Assistant Secretary or that meet the criteria in Appendix A [of CFATS] *must complete the CSAT Top Screen.*

The highest tier facilities, or Phase 1 facilities, are those specifically requested by the Assistant Security to complete the Top Screen; these are addressed by the Department first. All facilities that must complete the Top Screen are preliminarily tiered. These facilities are required to complete a Security Vulnerability Assessment (SVA), which provides more in-depth information that allows the Department to assign a final risk tier ranking to the facility. Preliminarily tier 1, 2, and 3 facilities must subsequently submit a CSAT Security Vulnerability Assessment. Tier 4 facilities may submit an Alternative Security Program (ASP) for the Department of Homeland Security to consider in accordance with 67 CFR 27.235(a). Tier 3 and 4 facilities may choose to submit an Alternative Security Plan for the Site Security Plan for consideration by the Department in accordance with 6 CFR 27.235(a)." [49]

Top screen questions:
http://www.dhs.gov/xlibrary/assets/chemsec_csattopscreenquestions.pdf
http://www.dhs.gov/xlibrary/assets/chemsec_csattopscreenusersmanual.pdf
Security Vulnerability Assessment (SVA) questions:
https://www.dhs.gov/sites/default/files/publications/chemicalsecurity_sva questions%20v3.pdf

APPENDIX D. POTENTIAL COVERAGE/OVERSIGHT OF WEST FERTILIZER BY DOT REGULATIONS

Ammonium nitrate is covered by DOT, according to the following paragraph taken from **§ 173.127**:

"173.127 Class 5, Division 5.1—Definition and assignment of packing groups. (a) Definition. For the purpose of this subchapter, oxidizer (Division 5.1) means a material that may, generally by yielding oxygen, cause or enhance the combustion of other materials. (1) A solid material is classed as a Division 5.1 material if, when tested in accordance with the UN Manual of Tests and Criteria (IBR, see § 171.7 of this subchapter), its mean burning time is less than or equal to the burning time of a 3:7 potassium bromate/cellulose mixture. (2) A liquid material is classed as a Division 5.1 material if, when tested in accordance with the UN Manual of Tests and Criteria, it spontaneously ignites or its mean time for a pressure rise from 690 kPa to 2070 kPa gauge is less then the time of a 1:1 nitric acid (65 percent)/cellulose mixture."[50]

All DOT requirements for ammonium nitrate are with regard to safe transportation. Last known inspection of the West Fertilizer site was conducted by DOT on September 23, 2011. The inspection resulted in a fine of

$5,250 with a total of 2 violations; illegible data on ASME placards and/or missing flammable gas placards (front and/or rear) and no security plan.[51,52,53]

All the penalties/fines were with regard to anhydrous ammonia.

"**Section 172.800(b)**[54] states, in part, "Each person who offers for transportation in commerce or transports in commerce one or more of the following hazardous materials must develop and adhere to a transportation security plan for hazardous materials that conforms to the requirements of this subpart. As used in this section, "large bulk quantity" refers to a quantity greater than 3,000 kg (6,614 pounds) for solids or 3,000 liters (792 gallons) for liquids and gases in a single packaging such as a cargo tank motor vehicle, portable tank, tank car, or other bulk container."

Section 172.802(b) [55] states

(a) "The security plan must include an assessment of transportation security risks for shipments of the hazardous materials listed in §172.800, including site-specific or location-specific risks associated with facilities at which the hazardous materials listed in **§172.800** are prepared for transportation, stored, or unloaded incidental to movement, and appropriate measures to address the assessed risks. Specific measures put into place by the plan may vary commensurate with the level of threat at a particular time. At a minimum, a security plan must include the following elements:
   (1) Personnel security. Measures to confirm information provided by job applicants hired for positions that involve access to and handling of the hazardous materials covered by the security plan. Such confirmation system must be consistent with applicable Federal and State laws and requirements concerning employment practices and individual privacy.
   (2) Unauthorized access. Measures to address the assessed risk that unauthorized persons may gain access to the hazardous materials covered by the security plan or transport conveyances being prepared for transportation of the hazardous materials covered by the security plan.
   (3) En route security. Measures to address the assessed security risks of shipments of hazardous materials covered by the security plan en route from origin to destination, including shipments stored incidental to movement.

(b) The security plan must also include the following:

(1) Identification by job title of the senior management official responsible for overall development and implementation of the security plan;
(2) Security duties for each position or department that is responsible for implementing the plan or a portion of the plan and the process of notifying employees when specific elements of the security plan must be implemented; and
(3) A plan for training hazmat employees in accordance with §172.704 (a)(4) and (a)(5) of this part.

(c) The security plan, including the transportation security risk assessment developed in accordance with paragraph (a) of this section, must be in writing and must be retained for as long as it remains in effect. The security plan must be reviewed at least annually and revised and/or updated as necessary to reflect changing circumstances. The most recent version of the security plan, or portions thereof, must be available to the employees who are responsible for implementing it, consistent with personnel security clearance or background investigation restrictions and a demonstrated need to know. When the security plan is updated or revised, all employees responsible for implementing it must be notified and all copies of the plan must be maintained as of the date of the most recent revision.

(d) Each person required to develop and implement a security plan in accordance with this subpart must maintain a copy of the security plan (or an electronic file thereof) that is accessible at, or through, its principal place of business and must make the security plan available upon request, at a reasonable time and location, to an authorized official of the Department of Transportation or the Department of Homeland Security."

Appendix E. Potential Coverage/Oversight of West Fertilizer by ATF Regulations

The Bureau of Alcohol, Tobacco, Firearms and Explosives (ATF) regulates ammonium nitrate- based blasting agents. It has regulations on the necessary distance to be maintained between ammonium nitrate and other explosive materials.

Subpart K – Storage§555.202. [56] Classes of explosive materials."(c) Blasting agents. (For example, ammonium nitrate-fuel oil and certain water-gels (see also § 555.11)."

§ 555.11 Blasting agent. [57] **"**Any material or mixture, consisting of fuel and oxidizer, that is intended for blasting and not otherwise defined as an explosive; if the finished product, as mixed for use or shipment, cannot be detonated by means of a number 8 test blasting cap when unconfined. A number 8 test blasting cap is one containing 2 grams of a mixture of 80 percent mercury fulminate and 20 percent potassium chlorate, or a blasting cap of equivalent strength.

An equivalent strength cap comprises 0.40–0.45 grams of PETN base charge pressed in an aluminum shell with bottom thickness not to exceed to 0.03 of an inch, to a specific gravity of not less than 1.4 g/cc., and primed with standard weights of primer depending on the manufacturer."

555.220 (1) [58] "This table specifies separation distances to prevent explosion of ammonium nitrate and ammonium nitrate-based blasting agents by propagation from nearby stores of high explosives or blasting agents referred to in the table as the "donor." Ammonium nitrate, by itself, is not considered to be a donor when applying this table. Ammonium nitrate, ammonium nitrate-fuel oil or combinations thereof are acceptors. If stores of ammonium nitrate are located within the sympathetic detonation distance of explosives or blasting agents, one-half the mass of the ammonium nitrate is to be included in the mass of the donor."

However, ATF does not regulate ammonium nitrate as fertilizer because of the exemption in subpart H.

Subpart H- Exemptions §555.141.(a).(8)[59] "Gasoline, **fertilizers**, propellant actuated devices, or propellant actuated industrial tools manufactured, imported, or distributed for their intended purposes."

If ammonium nitrate as fertilizer was covered by ATF, and stored nearby other explosives or other blasting agents, it would be required to be stored in accordance with the above table.

In the case of West Fertilizer, no other explosives are stored nearby to the best of our knowledge. Thus, even if ATF regulations had covered ammonium nitrate as fertilizer, the ammonium nitrate in the West Fertilizer facility would not to be required to be stored in accordance with the above table.

$555.220 Table of separation distances of ammonium nitrate and blasting agents from explosives or blasting agents

Donor weight (pounds)		Minimum separation distance of acceptor from donor when barricaded (feet)		Minimum thickness of artificial barricades (inches)
Over	Not over	Ammonium nitrate	Blasting agent	
0	100	3	11	12
100	300	4	14	12
300	600	5	18	12
600	1,000	6	22	12
1,000	1,600	7	25	12
1,600	2,000	8	29	12
2,000	3,000	9	32	15
3,000	4,000	10	36	15
4,000	6,000	11	40	15
6,000	8,000	12	43	20
8,000	10,000	13	47	20
10,000	12,000	14	50	20
12,000	16,000	15	54	25
16,000	20,000	16	58	25
20,000	25,000	18	65	25
25,000	30,000	19	68	30
30,000	35,000	20	72	30
35,000	40,000	21	76	30
40,000	45,000	22	79	35
45,000	50,000	23	83	35
50,000	55,000	24	86	35
55,000	60,000	25	90	35
60,000	70,000	26	94	40
70,000	80,000	28	101	40
80,000	90,000	30	108	40
90,000	100,000	32	115	40
100,000	120,000	34	122	50
120,000	140,000	37	133	50
140,000	160,000	40	144	50
160,000	180,000	44	158	50
180,000	200,000	48	173	50
200,000	220,000	52	187	60
220,000	250,000	56	202	60
250,000	275,000	60	216	60
275,000	300,000	64	230	60

National Fire Protection Association (NFPA) Official Standard No. 492, 1968.

End Notes

[1] http://co.williams.com/williams/news-media/geismar-update/

[2] http://www.nola.com/traffic/baton-rouge/index.ssf/2013/06/explosion_at_williams_olefins.html

[3] http://www.deq.louisiana.gov/portal/WilliamsOlefins.aspx

[4] Emergency Response Division, Office of Response and Restoration, National Ocean Service, National Oceanic and Atmospheric Administration, US Department of Commerce. http://incidentnews.noaa.gov/incident/8613

[5] 29 CFR 1910.109(i))

[6] Id.

[7] Shea, D.A., Schierow, L.J., Szymendera, S. (2013) Congressional Research Service. *Regulation of Fertilizers: Ammonium Nitrate and Anhydrous Ammonia.*

[8] 29 U.S.C. §654(a). http://www.gpo.gov/fdsys/pkg/USCODE-2010-title29/pdf/USCODE-2010-title29-chap15- sec654.pdf

[9] West Fertilizer Company Risk Management Plan, http://www.rtknet.org/db/rmp/rmp.php?facility_id= 100000135597&datype=T&reptype=f&detail=4&submit=GO

[10] *Fed. Reg. 6355, 6363 (Feb. 24, 1992)*

[11] http://www.chicagotribune.com/news/sns-rt-us-usa-explosion-regulationbre93k09h-20130421,0,7972342.story?page=2

[12] http://www.wfaa.com/news/texas-news/Documents-show-West--203543061.html

http://www.washingtonpost.com/blogs/wonkblog/wp/2013/04/18/the-texas-fertilizer-plant-explosion-is-horrific-but- how-common-is-this/

[13] http://keranews.org/post/epa-fined-west-fertilizer-plant-2006

[14] Inspection / Investigation Report No. 1220047. http://www.phmsa.dot.gov/staticfiles/PHMSA/DownloadableFiles/Press%20Releases/west_fertilizer_rpt_redact.pdf

[15] Compromise order. http://www.phmsa.dot.gov/staticfiles/PHMSA/DownloadableFiles/Hazmat/Enforcement/West%20Fertilizer%20Comp%20Order%20Jun%202012.pdf

[16] Notice of Probable Violation. http://www.phmsa.dot.gov/staticfiles/PHMSA/ Downloadable Files/Hazmat/Enforcement/West%20Fertilizer%20NOPV%20Jan2012.pdf

[17] http://opim.wharton.upenn.edu/risk/library/2001_JCB_3rdPartyAudits.pdf

[18] http://opim.wharton.upenn.edu/risk/downloads/archive/arch272.pdf

[19] Shea, D.A., Schierow, L.J., Szymendera, S. (2013) Congressional Research Service. *Regulation of Fertilizers: Ammonium Nitrate and Anhydrous Ammonia.*

[20] 29 U.S.C. §654(a). Available at: http://www.gpo.gov/fdsys/pkg/USCODE-2010-title29/pdf/USCODE-2010- title29-chap15-sec654.pdf

[21] 29 C.F.R. §1910.38(c). Available at: http://www.osha.gov/pls/oshaweb/owadisp.show_document?p_id=9726&p_table=STANDARDS

[22] West Fertilizer Company Risk Management Plan, available at http://www.rtknet.org/db/rmp/rmp.php?facility_id=100000135597&datype=T&reptype=f&detail=4&submit=GO

[23] 29 C.F.R. §1910.109. Available at: http://www.osha.gov/pls/oshaweb/owadisp.show_document?p_id=9755&p_table=STANDARDS

[24] 29 C.F.R. §1910.1200. http://www.osha.gov/pls/oshaweb/owadisp.show_document?p_table=standards&p_id=10099

[25] 29 C.F.R. §1910.119. Available at: http://www.osha.gov/pls/oshaweb/owadisp.show_document?p_table=STANDARDS&p_id=9760

[26] 29 C.F.R. §1910.119.

[27] Shea, D., Schierow, L. and Szymendera, S. (2013). Regulation of Fertilizers: Ammonium Nitrate and Anhydrous Ammonia. CRS Report for Congress. Available at: http://www.fas.org/sgp/crs/homesec/R43070.pdf

[28] How LEPCs and Other Local can include information from RMP in their ongoing work: http://www.epa.gov/oem/docs/chem/lepc-rmp.pdf

[29] What Does EPCRA Cover: http://www.epa.gov/oem/docs/chem/epcra.pdf

[30] Exemptions under Sections 311 and 312: http://www.epa.gov/osweroe1/content/epcra/epcra-qa_exempt_311.htm#s311e5_4

[31] http://emergencymanagement.supportportal.com/link/portal/23002/23016/Article/13919/Are-farm-suppliers-and- retailers-exempt-from-311-and-312

[32] Clean Air Act Section 112(r): Accidental release prevention/RMP Rule: http://www.epa.gov/osweroe1/docs/chem/caa112_rmp_factsheet.pdf

[33] Based on CAA Section 112(r)(2)(A): An accidental release is defined as "an unanticipated emission...into the ambient air from a stationary source."

[34] EPA list of regulated substances and thresholds: http://www.ncair.org/112r/files/40cfr68(9&68)_01141994.pdf

[35] EPA list of regulated substances and thresholds (pag 19): http://www.ncair.org/112r/files/40cfr68(9&68)_01141994.pdf

[36] RMP rule amendments: http://www.epa.gov/R5Super/cepps/pdfs/applicability-faq-200405.pdf

[37] RMP hearing. March 1999: http://books.google.com/books?id=oyy7IP0X3lAC&pg=PA18&lpg=PA18&dq=rmp+include+explosives?&source=bl&ots=hPVRLfJ49y&sig=3lguj7tddGoZH6Y05IkyagjetwM&hl=en&sa=X&ei=rtrEUYDvE9KJrQGV3IGIDw&ved=0CEcQ6AEwAw#v=onepage&q=rmp%20include%20explosives%3F&f=false

[38] EPA list of regulated substances and thresholds: http://www.ncair.org/112r/files/40cfr68(9&68)_01141994.pdf

[39] http://www.epa.gov/R5Super/cepps/pdfs/rmp-listed-chemicals-200708.pdf

[40] RMP requirements: http://www.epa.gov/emergencies/docs/chem/clean_air_guidance.pdf

[41] APPENDIX A. 40 CFR 68(page 9): http://www.epa.gov/osweroe1/docs/chem/Appendix-A-final.pdf

[42] http://www.epa.gov/osweroe1/docs/chem/ammonitr.pdf

[43] Decision rules on determining EPA Program level: http://www.epa.gov/osweroe1/docs/chem/Chap-02-final.pdf

[44] 2008 Consolidated Appropriations Act, Subtitle J, Secure Handling of Ammonium Nitrate ("Section 563", Public Law 110–161) http://www.gpo.gov/fdsys/pkg/PLAW-110publ161/pdf/PLAW-110publ161.pdf

[45] http://www.dhs.gov/ammonium-nitrate-security-statutes-and-regulations

[46] 72 Federal Register 17688-17745 (April 9, 2007) at 17697 (in page 17697, it is the last sentence of "1. Definition of ``Chemical Facility or Facility''", right above "2. Multiple Owners and Operators"). http://www.gpo.gov/fdsys/pkg/FR-2007-04-09/html/E7-6363.htm

[47] DHS list of chemicals: http://www.dhs.gov/xlibrary/assets/chemsec_appendixa-chemicalofinterestlist.pdf

[48] 72 Federal Register 65396-65435 (November 20, 2007) at 65407, http://www.gpo.gov/fdsys/pkg/FR-2007-11-20/html/07-5585.htm

[49] http://www.dhs.gov/risk-chemical-facility-anti-terrorism-standards-cfats

[50] 49 CFR § 173.127. Available at: http://www.gpo.gov/fdsys/pkg/CFR-2010-title49-vol2/pdf/CFR-2010-title49-vol2-sec173-127.pdf

[51] Inspection / Investigation Report No. 1220047. Available at: http://www.phmsa.dot.gov/staticfiles/PHMSA/DownloadableFiles/Press%20Releases/west_fertilizer_rpt_redact.pdf

[52] Compromise order. Available at: http://www.phmsa.dot.gov/staticfiles/PHMSA/DownloadableFiles/Hazmat/Enforcement/West%20Fertilizer%20Comp%20Order%20Jun%202012.pdf

[53] Notice of Probable Violation. Available at: http://www.phmsa.dot.gov/staticfiles/PHMSA/DownloadableFiles/Hazmat/Enforcement/West%20Fertilizer%20NOPV%20Jan2012.pdf

[54] 49 CFR §172.800. Available at http://www.gpo.gov/fdsys/pkg/CFR-2011-title49-vol2/pdf/CFR-2011-title49-vol2-sec172-800.pdf

[55] 49 CFR §172.802b. Available at: http://www.gpo.gov/fdsys/pkg/CFR-2011-title49-vol2/pdf/CFR-2011-title49-vol2-sec172-800.pdf

[56] 555.202 ATF Federal Explosives Law and Regulations (2012) http://www.atf.gov/ files/publications/download/p/atf-p-5400-7.pdf

[57] 555.11 ATF Federal Explosives Law and Regulations(2012) http://www.atf.gov/files/publications/download/p/atf-p-5400-7.pdf

[58] 555.220 (1) ATF Federal Explosives Law and Regulations 2012 http://www.atf.gov/files/publications/download/p/atf-p-5400-7.pdf

[59] ATF Federal Explosives Law and Regulations (2012) http://www.atf.gov/files/publications/download/p/atf-p-5400-7.pdf

INDEX

C

D

E

F

G

O

P

T

U